MAKE A HOME FOR WILDLIFE

Creating Habitat on Your Land
Backyard to Many Acres

CHARLES FERGUS

STACKPOLE
BOOKS
Guilford, Connecticut

This book is dedicated to the women and men who today are creating and advocating for the habitats that wildlife needs to survive. Some of these people are landowners. Others have made conservation their life's work and are employed by state and federal agencies, colleges and universities, and nonprofit organizations. They give wildlife a voice in today's world and teach us how to nurture the wild creatures whose continued presence makes this land a fascinating and wondrous place to live.

Published by Stackpole Books
An imprint of The Rowman & Littlefield Publishing Group, Inc.
4501 Forbes Blvd., Ste. 200
Lanham, MD 20706
www.rowman.com

Distributed by NATIONAL BOOK NETWORK
800-462-6420

Photographs by the author unless otherwise credited

British Library Cataloguing in Publication Information available

Library of Congress Cataloging-in-Publication Data
Names: Fergus, Charles, author.
Title: Make a home for wildlife : creating habitat on your land : backyard to many acres / Charles Fergus.
Description: Guilford, Connecticut : Stackpole Books, [2019] | Includes index.
Identifiers: LCCN 2018040811 (print) | LCCN 2018045810 (ebook) | ISBN 9780811767606 (e-book) | ISBN 9780811737722 (pbk.)
Subjects: LCSH: Wildlife habitat improvement—East (U.S.) | Habitat (Ecology)—East (U.S.)
Classification: LCC QL84.2 (ebook) | LCC QL84.2 .F47 2019 (print) | DDC 577.0974—dc23
LC record available at https://lccn.loc.gov/2018040811

∞™ The paper used in this publication meets the minimum requirements of American National Standard for Information Sciences—Permanence of Paper for Printed Library Materials, ANSI/NISO Z39.48-1992.

Printed in the United States of America

CONTENTS

1

You Can Make a Difference

ON A BRIGHT SUMMER DAY, I stood quietly at the woods' edge and looked out on the two-acre field we call "the Upper Meadow."

Orange hawkweed and yellow hawkweed, white daisies, red clover, and yellow buttercups spangled the field's lush greenery. (From violets and wild strawberries in May to goldenrod and New England asters in September, there's always something blooming in the Upper Meadow.) The slick, blade-like leaves of milkweed reflected light from the sky; unopened hot-pink flower clusters dangled from the plants' tall stalks. Hawthorn shrubs and apple trees stood above the lower plants, their branches dotted with still-green fruit.

I raised my binoculars and scanned the field. Winks of tan, orange, white, and pale blue came from butterflies' wings. A blue-green dragonfly hovered, then darted sideways to snatch a smaller insect out of the air. In the past, I've spotted deer, wild turkeys, foxes, coyotes, and bears in the Upper Meadow, but none of those large animals were present that day.

The Upper Meadow at the Butternut Farm in northern Vermont, kept as old field habitat through periodic mowing. Burke Mountain stands in the distance.

Lowering my binoculars, I looked at those verdant acres. It wasn't hard to imagine meadow voles down at ground level, scuttling along on narrow runways through the weeds and grass, pausing and using their teeth to snip the plants' stems and gain access to the upper, more succulent leaves and seed heads. I pictured field sparrows sitting on nests built on the ground, or attached to low blackberry canes, patiently incubating eggs or bringing insects to feed their growing nestlings. Caterpillars nibbling on vegetation. Spiders catching flies. Frogs and toads snapping up beetles and crickets. Shrews hunting earthworms. Weasels chasing mice.

I took a few steps into the open on a path I keep mowed along the field's upper edge. I glimpsed a sudden movement to my right: A hawk launched itself out of a tree, flapped its wings, and went soaring across the field. Its medium size and black-and-white banded tail identified it as a broad-winged hawk; I had seen one a few days earlier in the woods nearby, where I figured a pair might have built a nest. Perhaps the broadwing had been watching the mowed path for incautious rodents and snakes.

As I strolled down the path, I thought of what I had done to renew and maintain this opening at the forest's edge.

In 2003, my wife and I and our son moved from central Pennsylvania to an old farm in northeastern Vermont, a region of the state known as the Northeast Kingdom. The Butternut Farm had been named for the many butternut trees growing there. The house was rundown but fixable. The big attraction for us was the land: 120 acres, most of it forested but with several fields totaling around 16 acres. We could cut firewood in the woods to heat the house and make hay in the fields for our horses.

The property also included two old fields that, through the process of natural forest succession, were well on their way to becoming woods. Each field was about two acres. One field lay at a low elevation, behind and below the house; the other was higher up, at the base of a wooded hillside. One of our new neighbors told us that when he started farming in the early 1980s, he cut hay in both fields but eventually gave that up as the fields, left unfertilized, produced less grass and more weeds. As the years passed, shrubs and small trees seeded in.

After we finished remodeling the house (actually, after we pretty much rebuilt it around the dwelling's original post-and-beam structure), I turned my attention to the two old fields. When exploring the rest of our land and looking at neighboring properties, I had discovered that most of the area was either forest of varying ages or open hayfields. There were few old fallow fields. Restoring and perpetuating those two old fields would add to the diversity of habitats in our area, something that would be good for wildlife. It would be good for us, because we like to see wildlife. Delaying the task would only make it harder since the trees that had come in were getting bigger every year. As a bonus, removing the trees from the upper field would restore a long-range view to the north and east. That's where I started working.

Red maples, black cherries, white ash, and quaking aspens had risen above the weeds and grass in the old field. The largest of the saplings were around 25 feet tall. Scattered among the young trees were several dozen thornapple shrubs and wild apple trees—I didn't know how many there were until I'd crisscrossed the field several times tying bright orange flagging to their crowns. I planned to save the thornapples and apple trees since their fruits are relished by wildlife.

Using a chainsaw, I began felling the hardwood trees. Some of them hung up in their neighbors' crowns, and I had to push and pull until they came down, or, working from the bottom up, cut their trunks into stove-length pieces. I dragged the smaller branches out of the field and into the woods. It was hard labor but also satisfying, as I could see progress after each work session.

A great spangled fritillary feeds on nectar from a milkweed flower. In old fields, meadows, and other grasslands, wildflowers are an important source of food for insects, in turn preyed upon by amphibians, reptiles, birds, and mammals.
Tom Berriman

After removing the taller trees, I trimmed their stumps as low to the ground as possible. Then I walked along behind a big orange DR brush mower—a gasoline-powered brush hog made here in Vermont—and cut down the thumb-thick sprouts that were pushing up all over the field; had the sprouts been allowed to keep growing, eventually they would have become trees.

Finally, I had the Upper Meadow the way I wanted it: an expanse of grasses and forbs (herbaceous flowering plants: think wildflowers and weeds) scattered with thornapple shrubs and apple trees, plus one good-sized crabapple that I found hidden among the saplings. The view that my work restored—of forested mountains standing above an undulant patchwork of woods and fields—is beautiful in all seasons. Almost every day, my wife and I hike or snowshoe through the Upper Meadow, where we stop and take in the scenery—and look for wildlife—before continuing the last quarter mile home.

The following year, I did much the same thing with the second old field, the one lower down and closer to our house. The soil in the Lower Meadow is damper than that in the Upper Meadow. There are fewer apple trees, but Juneberry, elderberry, and pagoda dogwood grew profusely in the Lower Meadow, along with shrub willows and tamaracks. For variety's sake, and taking advantage of the native shrubs and small trees that were present, I left the Lower Meadow in a somewhat shrubbier condition than the Upper Meadow. I also kept a few patches of trees as islands of older habitat.

LANDOWNER STORY

An Island of Habitat in a Sea of Farmland

Carl Graybill planted evergreen trees plus bare-root seedlings of quaking aspen trees and dogwood shrubs in a habitat patch extending from his lawn into an adjacent crop field.

CARL AND MARY GRAYBILL LIVE IN A HANDSOME NEW COLONIAL-STYLE HOUSE on five and a half acres near Annville in southeastern Pennsylvania. Their property is part of a small development sectioned off from a farm. Surrounding the development lies some of the most fertile farmland in Pennsylvania, where corn grows in long, straight rows interrupted only by roads, houses, and farms with tall silos.

Carl is the retired director of the Pennsylvania Game Commission's Bureau of Information and Education. He and Mary love watching wildlife—which was not very common in the intensively farmed landscape when the Graybills built and moved into their new home. They decided to do something about that by turning part of a farm field behind their house into a patch of shrubs and young forest.

They took the first step in 2005 by planting 64 six-foot-tall hemlocks, spruces, and firs in three rows as a 200-foot windbreak that also provided cover for wildlife, mainly birds. Then in 2011, Carl obtained 150 bare-root seedlings of quaking aspen, a fast-growing tree that thrives in many different types of soil, plus 150 silky dogwoods, which are native shrubs that produce berries eaten by a wide range of birds. He planted the seedlings in a nursery bed in his garden and let them grow for a year so they would develop healthy root systems to give them a better chance of surviving once they were transplanted into the farm field.

The next spring, Carl marked out a semicircle 200 feet across and extending about 150 feet from the windbreak of conifers out into the field, a part of the Graybills' lot that was currently being farmed but would now become a home for wildlife.

In the nursery bed, the aspens had grown to be around four feet tall. Using a shovel, Carl planted them roughly eight feet apart scattered throughout the center of the semicircle.

The dogwoods were also around four feet tall. "Before they put out leaves, I gave them a haircut about 18 inches above the ground using a hedge trimmer," Carl recalled, a trimming that would stimulate their growth once they broke dormancy. He planted the dogwoods at roughly eight-foot intervals around the perimeter of the aspen stand and mixed in some seedlings of gray-stemmed dogwood and winterberry for variety. He also added five six-foot-tall spruce trees to the mix, randomly spaced in the patch, which altogether covered about three-quarters of an acre. He watered the trees and shrubs several times that first year, but other than that, "I pretty much just stood back and watched them grow." Everything did well except for the winterberry, which needs a wetter environment than the Graybills' land provided.

I first saw Carl's habitat patch in 2013, when it was a year old. It didn't look like much: some spindly plants with a few green leaves standing above bare ground. However, Carl pointed out a pair of brown thrashers darting into the windbreak of evergreens and then flitting off to land in a red maple he had planted in the lawn. "This is the first year we've seen brown thrashers since we moved in," he said. "They're probably attracted to secure nesting sites in the conifers. I'm hoping they'll stick around, and I'm looking forward to seeing other birds show up as the aspens grow."

When I returned four years later, the patch looked utterly different. The aspens now soared up 15 to 20 feet, with hundreds of coin-shaped leaves shimmering in the breeze. The dogwoods were sprawling head-high clumps with dense foliage. In between the dogwoods grew coneflowers, black-eyed Susans, and milkweed—plants sprung from seeds Carl had gathered in the wild or bought from a local nursery. At the edge of the habitat patch, foxtails grew profusely. "The mourning doves love to eat their seeds," Carl said. Just beyond the foxtails was a curving row of corn—part of a veritable sea of corn that stretched toward a low, wooded ridge on the horizon.

"The aspens have done really well," Carl said. "A few years from now they'll be even taller, and their underground root systems will send up more stems, making more small trees. Each autumn, their leaves fall and decompose, enriching the soil."

Carl uses an herbicide applied with a two-gallon hand-sprayer to limit plant growth around the edge and in parts of the habitat patch. "I want to keep the ground fairly open because I like to come in here and see what's going on, and I don't want to pick up any deer ticks," which are pests that carry Lyme disease and other maladies and that lurk in high weeds and grass. "I think that applying a small amount of herbicide is all right if you use common sense and are careful where you spray."

As we stood at the edge of the patch, doves flew overhead in small bunches, and robins scolded from the tops of the aspens. "Dozens of birds use this patch of young forest," Carl said. "We see gray catbirds and the occasional chickadee, and those brown thrashers have hung around. Cardinals and purple grackles nest in the spruces and firs, along with doves and blue jays. We see goldfinches, bluebirds, house finches, and plenty of robins."

Six years after they were planted, aspens in the Graybills' habitat patch are 15 to 20 feet tall. The dogwoods are head-high and taller, producing fruit for local and migratory birds, plus providing nesting sites and hiding cover.

In autumn, migrating birds drop in to rest in the thick cover and feed on fruits that the shrubs produce. "The smaller birds attract predators," Carl said. "A Cooper's hawk checks out the patch every so often; I've seen him perched in the hemlocks. And I see the results of his hunting now and then—little piles of dove feathers on the ground.

"Except for one year, every autumn since the patch was established I've found where a buck deer has rubbed his antlers on the aspens. The first year, a buck rubbed half a dozen of the aspens and broke off several a couple feet above the ground, but I didn't care because I knew the trees would just send up more shoots from their roots. Seeing those buck rubs and deer tracks in the dirt made me feel good." Carl has found plenty of deer droppings in the habitat patch, as well as several scrapes (areas of pawed-up soil made by rutting bucks) and a deer bed on the ground after a recent snowfall. He plans to put up a trail camera to see what other kinds of animals may be using the patch.

"We had something really encouraging happen this summer," he told me. "I hadn't seen a box turtle in many years, and, as you would expect, there were none on our property. One day Mary found one on our driveway, and I took it out to the young forest and put it on the ground among the aspens."

Carl and Mary note that neighbors sometimes complain, in a fairly good-natured way, about all the noise the birds make. "Well, grackles *are* pretty loud birds," Carl admitted. And then there are the droppings: whitewash splashed on cars and house windows, and fecal sacs—the waste products of young birds still in the nest—that parent birds carry away and drop wherever they decide to, which can include driveways, patios, and decks. "I've had neighbors say, 'Way to go, Carl,' referring to the mess," Carl said. "I just smile."

He added: "A lot of small landowners probably have never thought about how they can easily improve wildlife diversity—how they can have a real impact on the kinds and numbers of wild animals on their property, even if it's part of a housing development. All they need to do is to plan out a project that's within their means, buy some shrub and tree seedlings, and have the time and the passion to plant them and do a small amount of maintenance.

"I started this patch of young forest because I enjoy wildlife, particularly birds, and I wanted to see more of them. Mary and I get a lot of satisfaction out of seeing all the birds and hearing them call in the spring and summer.

"And it's not just the wildlife. In the fall, the aspens' leaves turn gold before they fall—they really stand out against the dark green of the conifers. The silky dogwoods have red stems with blue berries, and their leaves turn wine red.

"It's a young, growing forest now," he said. "When the aspens get tall, with a trunk diameter of six to 10 inches, and no longer provide dense habitat, I'll probably cut them down. Same with the dogwoods. Then they'll resprout and grow back thick again. Before then, I figure I've got another 10 to 15 years of great habitat in the young forest stage. And we'll continue to enjoy the wildlife that this patch of habitat attracts."

Now, more than a decade later, I brush-hog the fields every several years, weaving with the DR between the apple trees and the thornapples, cutting back the "whips"—the new little trees that continually push up from hidden roots—to keep the fields in an in-between state, one that is neither forest nor grassland, a type of habitat that is becoming increasingly rare in northern New England and throughout eastern North America.

The fields are wildlife hotspots. In autumn, bears feast on apples, sometimes breaking branches in the misshapen trees to get at the succulent fruit. Grouse peck at windfalls beneath the apple trees and clamber around in the thornapples eating the small red pomes in fall and early winter. Tracks in the snow show that fishers, foxes, and bobcats hunt in both openings, and a mink habitually follows the stream that trickles through the Lower Meadow. In the Upper Meadow one morning, I startled a barred owl that had just killed a robin (unlike most owls, barred owls sometimes hunt during daylight), the owl lifting up from the ground on soundless wings, its prey clutched in its talons. Kingbirds, indigo buntings, goldfinches, dark-eyed juncos, and cedar waxwings feed on insects, seeds, and fruits. Hummingbirds and monarch butterflies sip nectar from flowers, and the monarchs' caterpillars feed on the milkweed plants. Birds that breed in the nearby forest, including hermit thrushes, scarlet tanagers, and red-breasted grosbeaks, venture into the fields and catch insects to feed themselves and their young. In late summer, cedar waxwings pick the elderberry shrubs bare. In autumn, migrating birds drop in to eat the fruits of dogwoods and Virginia creeper, replenishing their fat supplies for the long journey south.

The Butternut Farm is a naturalist's paradise. It has wetlands, hardwood forest, stands of evergreen trees, grass fields, and weedy old fields. Adding to the mix are vernal ponds, dead snags, rotting logs, rock ledges, and old stone walls. Just down the hill, on a neighboring property, is a cedar swamp (if you want to get technical about it, a "calcareous fen"). It's a rare occasion when my wife and I do not see interesting animals, or signs of their presence, when we venture outdoors.

But you don't have to own a 120-acre hill farm in Vermont to shape and manage your property in ways that will let you help—and enjoy viewing—our region's native wildlife.

Whether you live on a lot in town, own an acre or two in the suburbs, spend your weekends at a 10-acre woodland retreat, manage a working forest or farm, or belong to a hunting club with hundreds or even thousands of acres—you can make changes to the land that will transform your property into a better home for wildlife. Habitat projects can be simple or complex; they can be short-term or spanning decades. Their cost can be minimal (a few hours doing pleasant work outdoors, along with some gas to run a chainsaw or a brush hog) or run into the hundreds or even thousands of dollars. State and federal conservation agencies and nonprofit wildlife organizations offer advice and funding to landowners who want to make habitat in various ways. In some cases, creating habitat—for example, by conducting a timber harvest to add age diversity to a woodland—can generate income that can then be used to buy more land or make additional improvements for wildlife.

Make a Home for Wildlife focuses on eastern North America from southern Canada south to northern Florida and west to the Great Plains. In general, this area receives ample rainfall that supports the growth of a wide variety of plants. Those plants, in turn, provide sustenance for creatures from tiny invertebrates such as mites and springtails all the way up the food chain to large browsing mammals such as deer and moose, as well as predators such as coyotes and black bears.

Most of the land in the East is privately owned, including the forested land that provides most of the habitat for wildlife today. According to the U.S. Forest Service, more than half of the forest-

Black bears eat clover, grass, and other meadow plants and find prey in grassy openings. In areas that are mainly forested, grasslands are wildlife hotspots.

land in the United States—some 441 million acres, most of it in the East—is owned and managed by around 11 million private ownerships. Of those private ownerships, the Forest Service classifies 95 percent as "Family and Individual," as opposed to corporate, owners. Here in Vermont, 87 percent of the land is privately owned. In Pennsylvania, where I grew up and lived for many years, that figure is 84 percent. Pennsylvania has an estimated 738,000 private forest landowners; a quarter of the Keystone State's privately owned forest exists in parcels smaller than 20 acres, and more than half of all private forest holdings in Pennsylvania (some 420,000 of them) are 10 acres or smaller.

Individuals also own other key habitats throughout the East, in the form of grasslands, shrublands, and wetlands. People's backyards, and larger lots around houses sited on former farmland or in wooded areas, also have the potential to provide important food and cover for wildlife.

State and federal agencies can manage habitat on state and national forests, wildlife management areas, and wildlife refuges. However, because most land in the East is privately owned, it falls on private landowners—plus members of land trusts, folks who oversee nature preserves and belong to hunting clubs, and citizens who sit on conservation commissions for municipalities that own

Many animals eat apples from wild apple trees. Landowners can cut down taller trees that cast shade on apples to restore the fruit trees' productivity.

land—to collectively make enough good-quality habitat so that wildlife can thrive. The benefits are many, both for the animals and for us as humans. In these increasingly busy and technology-oriented times, it's important for us and our children to be able to see wildlife to support our mental health and well-being, learn about the natural world, and have experiences that will lead us to value and take care of the land.

Make a Home for Wildlife provides information about animals and their habitat needs. It describes four basic habitat types: forests, shrublands, grasslands, and wetlands. It covers features that add diversity to a piece of land and make it more attractive to wildlife—things such as den trees and snags, streams and ponds, openings in the forest, brushy fencerows, nut-bearing trees, sources of fruits and berries, and more.

This book explains different approaches that can be used to make, refresh, and maintain a variety of habitats for wildlife. It is less about prescribing specific habitat management techniques than about giving landowners the tools—in the form of knowledge about wildlife, plants, and habitats, along with knowing how to find resources and obtain professional guidance—that will let them set realistic goals for creating and improving habitat and then following through on projects. I hope there's enough information in these pages to get you started. It's very helpful if you know how to use the internet, as an incredible amount of practical, detailed information can be found there.

Scattered throughout the book are "Landowner Stories" about people who have created or improved habitat in different states and regions and in different ways, as well as the satisfaction they've gained from seeing and encountering wildlife—and, in some cases, helping species whose numbers have fallen to dangerously low levels. I've also included some "Wildlife Sketches," stories about wildlife and how animals use the habitats in which they live. I hope you enjoy them.

I also hope that *Make a Home for Wildlife* will help you see your property in new ways and then plan and carry out practices and improvements to make it more inviting to the creatures with which we share the land.

2

Wildlife and Habitat Basics

THE PATH LED THROUGH A STAND OF HICKORIES AND OAKS. I paused at the base of a medium-sized chestnut oak. A few years earlier I had nailed a box, made of scrap wood and with a two-inch entry hole, onto the tree's trunk about 10 feet up. I had put up the box for cavity-nesting birds, hoping to attract a pair of great crested flycatchers, a species I sometimes glimpsed in our Pennsylvania woods—active brown-and-yellow birds whose raucous *wheep, wheep* calls echoed from the forest's leafy canopy.

But what I saw that morning, clinging to the oak's deeply ridged bark just above the nest box, were flying squirrels. Four of them. They were youngsters, about half the size of adults. Their pearly gray coats glistened in the morning light, and their big, dark eyes seemed to look down at me more with curiosity than with fear.

A good habitat will supply an animal with food, such as the caterpillar clenched in this song sparrow's bill. Wildlife also needs water and cover within an area in which they can move about without exposing themselves to excessive danger.
Tom Berriman

The forest on our 30-acre property was about 60 years old, according to an elderly neighbor who had seen this section of Pennsylvania's Allegheny Front logged off in the 1930s. By the 1990s, when we were living there, the trees had grown back. Many of them were fairly large, but few were old or large enough to have developed the hollows in their limbs and trunks that great crested flycatchers, along with other forest birds such as screech owls and woodpeckers, use for nesting. So I built and put up half a dozen nest boxes. Flying squirrels hadn't entered my mind—but there they were, flattened against the oak's rough bark, staring down at what was no doubt the first human they had ever seen. Their mother had probably given birth to them inside the box, and now they were almost old enough to venture out from its security and begin exploring the habitat into which they'd been born.

An animal's habitat is the physical place where it lives. A good habitat provides a creature—whether a mammal, bird, reptile, amphibian, or insect—with the resources it needs to survive. In its habitat, an animal can grow from birth to adulthood, remain healthy, evade predators (at least for the most part), find shelter during harsh weather, and mate and reproduce.

Some animals are generalists, able to thrive in more than one type of habitat: White-tailed deer can live in large expanses of deep woods, in small woodlots, in thickets, in farmland, and in the suburbs and on the fringes of cities. Other creatures have more specialized habitat needs. The wood thrush, a migratory songbird with a beautiful flutelike song, usually nests in mature forest where tall trees stand above an understory of smaller trees, shrubs, and low plants, with abundant leaf litter on the ground.

A good habitat will supply an animal with food, water, and cover. A wild creature needs these essentials within a physical area, or space, in which it can move about without exposing itself to excessive danger. The blend of food, water, shelter, and space determine an area's *habitat suitability* for a species of wildlife or a suite of species.

FOOD

Based on what they eat, animals can be classified as *herbivores*, *carnivores*, and *omnivores*.

Herbivores eat plant matter, including buds, stems, shoots, sap, roots, bark, leaves, flowers, nectar, fruits, seeds, and nuts. Deer, moose, hares, rabbits, voles, porcupines, beavers, and muskrats are all herbivores. Some animals that mainly eat plants may also consume some animal protein. Mice, chipmunks, and squirrels (including cute little flying squirrels such as the ones born in the nest box on our old place in Pennsylvania) subsist mainly on nuts, seeds, berries, and buds, but they will also devour birds and their eggs or nestlings if given a chance. Even deer will eat some animal-based protein (birds and eggs) if the opportunity arises.

Carnivores eat other animals. Otters, fishers, mink, weasels, and shrews are carnivores whose diets include both terrestrial (land) and aquatic (water) animals. Bobcats and lynx are carnivores. Hawks, falcons, and owls catch and consume smaller creatures. Vultures eat carrion, the flesh of dead animals, and they also prey on live animals that they can catch and overpower. Herons eat fish, crustaceans, mice, shrews, and other small mammals. Woodcock feed on worms; whip-poorwills vacuum up flying insects; woodpeckers excavate insect grubs from the wood of trees; and songbirds snap up insects of many different types and life stages. Snakes eat small mammals, birds, other reptiles, amphibians, and insects. Toads and frogs consume huge numbers of insects and other invertebrates.

Omnivores eat both plant and animal foods. Black bears are highly omnivorous: In spring, they hunt down and eat white-tailed deer fawns, and they also dine on carrion, garbage, birdseed, and plants ranging from skunk cabbage to the new leaves of trees. In summer, they eat an assort-

Bobcats are carnivores—meat-eaters—found in much of the East. A bobcat's home range in summer may take in only a few square miles, but its territory may expand in winter when prey becomes scarce and the cat broadens its search for food.
© *iStock.com/gatito33*

ment of animal foods, including mammals, birds, frogs, crayfish, centipedes, and insects, along with grass, mushrooms, and berries. In autumn, they build up fat for winter's hibernation by gorging on high-energy foods such as acorns, hickory nuts, and beechnuts. Raccoons, opossums, and skunks are omnivores. Coyotes and foxes, although mainly carnivorous, also eat fruits and berries. Many birds switch between insects and fruits, depending on the season and food availability in a local area.

WATER

Animals drink from rivers, streams, wetlands, springs, ponds, and puddles. Some sip dew, and others get the moisture they need from their food. Certain kinds of wildlife live only in or near water. Amphibians—frogs, toads, and salamanders—need water for breeding. Many aquatic plants provide important food for wildlife. Spring seeps, where groundwater wells up to the surface of the land, often stay snow-free during winter; animals can drink there, and plants and insects in such places are key food items when the ground is frozen and snow lies deep. Low areas along streams and rivers, called *riparian zones*, often have damp, rich soils that support lush plant life that gives rise to abundant food and cover.

COVER

Think of cover as screening or protection provided by a habitat. Wild animals use it for traveling, resting, nesting, rearing young, avoiding predators, and sheltering from extremes of

Turtle Time

Wood turtles can live more than 30 years in the wild. Counting the number of annual lines on each scute (the raised plates that link together to form the turtle's upper shell) reveals the turtle's age.
Jonathan Mays

ON A HUMID SUMMER DAY, HIKING ALONG A TRAIL at Pine Ridge State Forest Natural Area in southern Pennsylvania, I met a turtle. A wood turtle. It seemed the turtle had seen me first, or had otherwise sensed my presence, for it had withdrawn its head, legs, and tail inside its shell. Its soot-colored face peered out from beneath the protecting carapace.

I sat down beside the turtle. I had plenty of time and decided to spend some in the company of this reptile.

Over the next five minutes, the turtle extended its head out of its shell, millimeter by millimeter. Reddish orange markings showed on the turtle's neck. The turtle appeared to fix a dark, shiny eye on me. I wondered what that eye saw, and I wondered what the turtle thought about what its eye registered.

The turtle's upper shell was composed of raised pyramidal plates. The shell was a rich dusky brown; darker radiating lines marked each plate, of which there were

38 that were knit tightly together. Each plate, or scute, was made up of a series of concentric quadrangles, much like the elevation lines on a topographic map. Each line signified a year's growth. When I got my face down near one of the scutes—causing the turtle to quickly pull back its head again—I could make out approximately 25 growth lines. Twenty-five years; a quarter of a century. A generous life span for a wild animal. Even the larger mammals, such as deer and bears, rarely live longer than 10 or a dozen years. And wood turtles can live to be older than that: Wild ones have survived more than 33 years, and captives have lived past 50.

This wood turtle's shell was about eight inches long by five inches wide. On each side, around the middle part of the carapace, the shell was worn smooth, from the turtle pushing itself through weeds and brush. Taking the shell in my hand, I lifted the turtle off the ground. In contrast with its earth-colored carapace, the bottom of the turtle's shell—the plastron—was a rich straw-yellow with dark mahogany blotches.

I set the turtle back down.

After quite some time, the turtle brought its head out again. Then it partially extended its limbs; similar to the head and neck, these also had reddish orange markings, and they were tipped with sturdy black claws. The turtle did not walk off. It continued to look at me, or at least to look in my general direction.

More time passed. I found myself listening to crows cawing and a wood thrush singing. Insects buzzed and chittered. The breeze whispered through the crowns of the forest trees. It was all rather mesmerizing and quite pleasant. At times, I would forgot about the turtle sitting next to me. Then I would come back to myself and look at the turtle again. It did not move. After a while, I realized that I was becoming a little impatient for the turtle to do something. To do anything. The loose skin sheathing the turtle's neck pulsed faintly in time to the beating of its heart. A fly landed on the turtle's slatey, flat-topped head. After a few minutes, the fly walked across the turtle's eye, at which point the turtle actually blinked.

When I'd first sat down, the sun had been directly overhead. Now the trees cast long shadows across the path. I got up, stretched, and brushed the dust off my pants. The turtle reposed in the same spot where I had put it down after examining the bottom of its shell. It stayed there as I walked away down the trail.

The turtle, of course, did not exist to entertain or to please me. It did not grasp the concept of "pleasing" anybody or anything, except perhaps its own reptilian self. Yet, in its placid, unassuming manner of existing—of simply being—it had done just that. It had sent me a subtle message, telling me to slow down, absorb my surroundings, and become present and attentive to the sights, sounds, and smells of the natural world through which I was privileged to pass.

Cottontail rabbits are prey for hawks, owls, foxes, and other animals. They need thick cover to escape from predators, but even in a good habitat, a rabbit will rarely live longer than a year.
Victor Young, NHFG

weather, including heat, cold, and storms. Forest trees offer cover, both when they are young and growing together closely in dense stands, and also when they are mature. Shrubs and grass offer cover. Biologists often refer to the composition and arrangement of vegetation in a habitat as its *structure*.

A brush pile built at the edge of a field can be an important cover feature—a hideout into which a cottontail rabbit can dart to escape a fox. A rotting log, a hollow space in a tree's trunk (or a nest box put up by a landowner), dense vegetation bordering a stream or pond, a cave or a rock overhang, a burrow dug into a hillside—all represent types of cover used by wildlife. Many animals use different kinds of cover at different times of the year. The dense branches and foliage of evergreen trees provide secure nesting sites for birds in spring and summer; they also blunt winter winds and offer *thermal cover* where animals can rest without having to burn excessive calories during cold and snowy weather.

SPACE

Every animal needs enough space in which to meet its food, water, and cover needs and to interact with other members of its own species, including potential mates. This zone of personal space is known as an animal's *territory* or *home range*. Some creatures, especially predators, have large home ranges. Coyotes, almost ubiquitous across eastern North America, may range over eight to 16 square miles. (If it were square, a 16-square-mile territory would be four miles on each side.) A bobcat's range in summer may be only a few square miles, but its territory may expand to 20 to 40 square miles in winter when the cat broadens its search for food. Black bears' territories vary depending on an individual's sex and the quality and abundance of food and cover in a given area. Female bears with cubs may range over six to 19 square miles, while males may cover 60 to 100 square miles, their home ranges overlapping with those of several females. Bears' ranges will be smaller in areas where food sources and cover abound.

Small animals tend to have smaller ranges than large animals. A red-backed vole may live in a thickly vegetated forest opening a quarter acre in size; a chipmunk (larger than a vole) might use half an acre; a red squirrel may need an acre of woods to find food and shelter year-round. Home ranges of birds also tend to correlate with body size and weight: Most wood warblers, which are quite small, have a home range of five to 15 acres; woodpeckers, up to 200 acres; and ravens and goshawks may cover 10 square miles. A red-backed salamander may live out its life in the soil, rotting wood, and leaf litter in an area that could be covered by a beach towel.

GENERAL HABITAT TYPES

Forests

Eastern North America is rich in forests. More than 600 kinds of trees and shrubs grow in the eastern half of our continent. They include ones with needle-like foliage as well as broad-leaved types whose leaves are flattened structures of different shapes and sizes, depending on the species and age of the tree. Pines, hemlocks, spruces, and firs are common needle-leaved trees; they are also called *evergreens*, because they hold onto their green needles year-round, and *conifers*, because they produce seeds borne in woody cones. Broad-leaved trees include oaks, maples, birches, beeches, hickories, and many more. They are classified as *deciduous* (from the Latin *decidere*, meaning "fall down or off"), since they drop their leaves in autumn each year to conserve water and protect against damage from snow and ice buildup. Many wooded areas have a mix of both evergreen and deciduous trees.

Forests can be young, middle-aged, or old. Because a large percentage of the East is forested, making your land attractive to wildlife may involve managing a wooded tract by favoring certain kinds of trees over others, letting trees grow and mature, or harvesting trees so that a portion of your woods will be set back to a younger growth stage. Such actions will affect the types of wildlife a forested property will attract and support. (See Chapter 4, "Forests for Wildlife," for more information.)

Grasslands

In many parts of the East, most grassy areas represent temporary openings in what has historically been a largely forested landscape. And the forest always wants to come back. If a field ceases to be cut for hay, pastured by livestock, or periodically swept by fire (scorching and killing invading trees or shrubs), it will gradually become shrubland and then woodland. Centuries in the past, grasslands of varying sizes broke up the forest, created by disturbances such as fires, Native Americans' land-clearing to grow crops, and flooding caused by spring snowmelt and extensive networks of beaver dams. In the South, grasslands and savannas were more common than in northern areas, but even in heavily wooded regions, such as New England, there were sizable openings and numerous smaller grasslands interrupting the forest. These openings were among the first areas to be settled by English and European colonists.

Today, grasslands of varying qualities and usefulness to wildlife exist on working and recently abandoned farms, along roads and railroad rights-of-way, in old cemeteries, and in fields planted by private landowners and by conservation agencies and organizations. Corporate parks, airports, recreation fields, capped landfills, and reclaimed strip mines are also grassland habitats. Grasslands can support native grasses, sometimes called *warm-season grasses*, or non-native grasses, also known as *cool-season grasses*. Wildflowers and weeds often mix with the grasses. Grasslands offer cover to wildlife, and they represent food hotspots since their greenery provides food to herbivores as well as large numbers of insects, which in turn are preyed on by birds and other animals. (See Chapter 5, "Grasslands for Wildlife," for more details.)

Shrublands

These valuable habitats are neither grasslands nor forests but rather a combination of both. Shrublands are relatively open areas scattered or overgrown with woody shrubs typically less than 10 feet tall and often mixed with some small trees. Grasses, sedges, wildflowers, and other low plants grow between the shrubs and saplings. Shrublands arise on old farmland abandoned within the last 5 to 40 years. The shrubs that spring up in such settings may be either native to the East, or aggressive, non-native types called *invasives*. (Both furnish food and cover to wildlife, but the native shrubs provide more-nutritious food for the animals that coevolved with them.) Shrublands also persist on sites periodically swept by fire and in places where soils or wet conditions limit tree growth. We have lost many shrubland habitats in recent decades as old fields have been developed and as trees on such sites have grown taller and shaded out the shrubs and other low plants, turning the habitat into a forest. Other names for shrublands include *thickets, early successional habitat*, and *shrub-sapling openings*. (Learn more in Chapter 6, "Shrublands for Wildlife.")

Wetlands

Eastern wetlands include bogs, fens, swamps, beaver ponds, hardwood bottomlands, wet meadows, potholes, sloughs, groundwater seeps, pocosins, and Carolina and Delmarva bays. Wetlands such as vernal pools, where frogs and salamanders gather to breed in springtime, are very important to the *biomass* (the total mass of living organisms) in an area. Ecologists estimate that more than half of all wetlands in the lower 48 United States have been lost to draining, paving, and agricultural activities over the last 200 years. Sometimes called our most endangered habitat, wetlands store and slowly release rainwater, reducing erosion and flooding, and they purify water by filtering out sediments, excess nutrients, and pollutants. They also provide high-quality habitat for a range of wildlife. Landowners can restore wetlands that were drained in the past; they can also improve the quality of existing wetlands and create new wetlands on both damp and dry sites. (Chapter 7, "Wetlands for Wildlife," has more information.)

Wetlands and ponds are excellent habitats for wildlife. Landowners can restore wetlands that were drained in the past, improve existing wetlands, create new wetlands, and build permanent and temporary ponds.
© *iStock.com/ChuckSchugPhotography*

SPECIAL HABITAT FEATURES

Features that add variety and attract wildlife include ponds, streams, spring seeps, rock ledges, deer wintering areas, fencerows, den trees, snags, dead and down wood, nut- and fruit-bearing trees, and rare plant communities. Landowners can protect, enhance, and create many of these assets. (Chapter 8, "Special Habitat Features," gives details.)

KEY CONCEPTS

University extension personnel, biologists with conservation agencies, and professional foresters often advise landowners on how to improve their properties for wildlife. The following terms and concepts will help you understand resources these specialists offer, including pamphlets, fact sheets, instructional booklets, videos, webinars, and best-management-practices guides, most of which are available for free or for a nominal cost or can be viewed on or downloaded from internet websites. You may want to skim this section now and refer back to it later as you explore ways of turning your property into a better home for wildlife.

An *ecosystem* is a community of living organisms that interact with each other and with nonliving parts of their environment, including the air, soil, and water. *Organisms* include both plants and animals. *Biodiversity* refers to the variety of different life forms found in ecosystems and in nature. The more variety and complexity in a habitat—or in a group of nearby habitats—the more kinds and greater numbers of wildlife will live there.

A *niche* refers to the special place an animal occupies in a habitat, and, more broadly, how it fits into an ecosystem. For instance, when several different species of warblers live in the same woodland, they will usually occupy different feeding and nesting niches to avoid competing directly with one another for limited resources.

Plant succession takes place when one community of plants replaces another community over time. A classic example is when an old field or a grassland becomes a shrubland with scattered tree saplings and when, after decades, the trees increase in height and size, shade out the shrubs and low plants and the former field becomes a forest. (This sequence is also called *forest succession*.)

Animals of open habitats, including grasslands, are sometimes called *early successional* species. Shrublands, thickets, and areas of young forest support both *early and mid-successional* wildlife. Mature forest meets the habitat requirements of *late-successional* species. By taking management actions, landowners can "set back" succession to favor animals that need younger, thicker habitats. Or they can allow the successional process to continue, which will benefit late-successional wildlife. A high-quality habitat for one kind of wildlife may be a poor habitat for another.

Natural disturbances include flooding caused by beavers' dam-building activities and by spring snowmelt or heavy rain, wildfires, ice storms, high winds spawned by thunderstorm microbursts and hurricanes, insect damage, and plant disease outbreaks. Humans have blunted some of these forces: We suppress wildfires, limit beaver populations and activities, and build flood-control dams. Landowners can attract more and different kinds of wildlife by carefully applying certain habitat management techniques to mimic natural disturbances, thereby adding diversity in age, species, and density of the shrubs, trees, and other plants that offer food and cover.

Vertical structure or *stratification* refers to how plants are layered in a habitat. For example, a healthy forest will have a ground layer, a shrub layer, medium-sized trees in the *understory* or *mid-canopy*, and tall trees whose upper limbs form the *forest canopy*. The more layers a habitat has, the more attractive it will be to wildlife.

Habitat patchiness or *interspersion* describes the way a habitat is put together on a horizontal scale. For example, two types of vegetation, such as grasses and other low plants that supply food,

Many kinds of wildlife eat wild grapes, the fruit of sun-loving vines that grow in wooded areas, thickets, and edge habitats.

may combine with clumps of shrubs that offer hiding cover. Conservationists sometimes refer to habitat patchiness as *horizontal structural diversity*.

Edge habitat exists where two different types of plant growth meet, such as a forest and a field or a grassland and a wetland. Scientists refer to these areas as *ecotones*. Edges can be good for many kinds of wildlife but not for all.

Fragmentation occurs when large blocks of habitat are broken up by roads, housing, intensive farming, gas-well drilling pads, and other human developments. Fragmentation can lead to animals being killed when they cross roads or are forced to travel long distances to find food, water, cover, or mates. Fragmentation encourages predators such as raccoons. It also makes it easier for the brown-headed cowbird, a parasitic species, to lay its eggs in other birds' nests.

Corridors are linear patches of cover that animals use to move between separate habitat areas to find food or water or meet unrelated individuals for mating. Corridors let wildlife *disperse* and colonize new areas of habitat that have been created by plant succession, natural disturbances, or humans' land-management activities. Corridors can be on dry land; mosaics of wetlands can also act as corridors.

Natural resource specialists often use the term *invasives* to describe non-native trees, shrubs, and low plants that grow and spread aggressively, outcompeting native plants that offer better food to wildlife. Certain insects, such as the gypsy moth and the emerald ash borer, are invasive species. Wild hogs are an example of invasive mammals. Native species can be invasive in some conditions, such as when a heavy growth of ferns or beeches prevents other plants or trees from growing on a site.

An *umbrella species* or *guild species* is a type of animal whose habitat is used by many other kinds of wildlife. Making habitat for an umbrella species automatically helps a large group or *guild* of different animals.

Biological carrying capacity is the number of individuals of a certain species that an area can support in good health without the habitat being damaged. Biological carrying capacity will change through the seasons as the productivity of key food plants increases or decreases. For instance, in years when oaks produce a bountiful acorn crop, a forest's biological carrying capacity for mice, squirrels, deer, and bears will be higher than in years when the oaks produce few or no acorns.

Cultural carrying capacity is the number of individuals of a species that people are willing to tolerate in a given area. An area's cultural carrying capacity for beavers (which build dams, flood roads, and cut down trees) or deer (which browse ornamental shrubs and tree seedlings, and are part

Animals travel through brushy cover growing where forested habitats meet fallow or farmed fields. Landowners can create habitat corridors by cutting back woodland margins and letting young trees grow back, or by planting native trees or shrubs.

of the Lyme disease cycle) may be lower than the same area's biological carrying capacity for these animals.

Wildlife populations naturally rise and fall at different times of the year. Factors affecting long-term trends in local and regional populations include *mortality* from predation, hunting, accidents, disease, and changes in or loss of habitat. Enough good-quality habitat can help keep most animal populations stable. Creating specific kinds of habitat can increase the numbers of certain types of wildlife that a landowner may wish to encourage—songbirds, waterfowl, and game birds or mammals, butterflies, or salamanders.

3

Get to Know Your Land and Explore Its Potential

PROPERTIES COME IN MANY DIFFERENT SIZES AND SHAPES. Fortunately, there's almost always something that can be done on any plot of land to help wildlife—which will ultimately make your own time outdoors more interesting and enjoyable. If your property is small, such as a building lot on which your house sits, you probably already know what sorts of plants grow on it, and you may have a good idea of what types of wildlife frequent it or pass through. However, for parcels larger than an acre or so—and especially ones 10 acres and up—there are important early steps you should take before planning and launching a habitat-improvement project.

What do you have on your property? Hayfields? Woods? Wetlands? Learn which types of wildlife use your land. It may make sense to improve existing habitats rather than create new ones from scratch. You can even make important habitat in your backyard.

Steve Capel is a biologist who has worked for the Virginia Department of Game and Inland Fisheries and the Wildlife Management Institute. "Folks who want to help and to see more wildlife may think, 'I want to attract hummingbirds,' or 'I want to increase deer numbers,'" he says. "That's probably not the correct first step. They need to give thought to what their property already has and what's nearby. There may be no need to duplicate a type of habitat that's already abundant in a general area." He continues, "It's actually easier, and I believe it's ultimately a lot more satisfying, to characterize the different habitats you have and then to improve and optimize one or more of those habitats rather than trying to make a whole new habitat on your land."

Dr. Margaret Brittingham, an extension wildlife specialist at Pennsylvania State University, agrees. "The first thing a landowner should do is really get to know their land," she says. Brittingham conducts field research on different types of wildlife, especially songbirds, and creates landowner-oriented publications explaining how folks can attract and help animals. She also brings a landowner's perspective to her work: She and her husband own a 50-acre property in a rural valley north of Penn State. There they are developing a forest management plan aimed at helping wood thrushes, scarlet tanagers, and other migratory songbirds that breed in eastern woodlands during spring and summer and then winter in Central and South America—species whose populations have generally fallen in recent decades.

"A lot of folks don't have an in-depth understanding of what they own," Brittingham says. "That can lead them to have unrealistic ideas about how to create or improve habitat. Let's say you own 10 acres of forest on the side of a wooded mountain and decide that what you'd really like to do is encourage grassland birds such as meadowlarks or bobolinks. That's not a realistic goal for that particular property. It's doubtful that it could be reached, which could lead to a lot of frustration and wasted effort." This is true because creating a grassland on such a site would be difficult and expensive, and if and when it would become established, it's unlikely that grassland birds would find and use the habitat.

Instead, Brittingham advocates coming up with a clear picture of what you own and then figuring out how to make that habitat as good as it can be for the kinds of wildlife that currently use it—or that may be passing it by, if it doesn't supply the food and cover they need.

Brittingham advises landowners to answer some basic questions about their land: "Is part or all of it forested? Does it have any grassy openings or fields? Are there any streams or low damp places or wetlands? Are there areas overgrown with shrubs, and are those shrubs native to the region or are they invasive alien types? Is the land flat or sloping? Is the soil fertile or poor?" After answering these and some related questions, landowners will be much better able to develop achievable goals for creating or restoring wildlife habitat.

Next, "collect a library of field guides, natural history books, and publications about wildlife and different kinds of habitat," Brittingham says. Pertinent publications on animals and their habitat needs can be obtained for free or for a nominal cost from state and federal wildlife and forestry agencies, nongovernmental organizations such as the Audubon Society, the Nature Conservancy, and the National Wildlife Federation, and cooperative extension services supported by state land-grant universities. Searching the internet is a good way to locate these and many other resources. Get to know your state's extension service by checking out their website and phoning or visiting your county agent, whose number will likely be in your telephone directory. State extension services are incredible sources of information and advice for those wanting to help wildlife.

Biding Time

Pin cherries are short-lived trees that grow in direct sunlight. Their fruits feed birds, which then deposit the trees' seeds in their droppings, letting the cherries spread across the landscape. The seeds can remain viable for 50 or more years.
Donna Kausen

THE TRIO OF LANKY BRONZE-BACKED TURKEYS FROZE for an instant before vacating the trail and legging it off through the woods. I walked up to where they'd been scratching and pecking, and I scanned the ground.

The fruits lay like rubies on the brown fallen leaves. I picked one up: rounded, a quarter-inch in diameter, with a thin skin covering juicy flesh. When I tasted the fruit, my mouth puckered. Beneath the sour flesh lay a tiny oblong stone. I looked up, to where a pin cherry tree arched its slender limbs.

A mature pin cherry is a smallish, unassuming tree that does not yield much fruit—about two-thirds of a quart in a good year—but taken together, all of the pin cherries in a forest stand can provide a late-summer bounty. Robins, thrushes, jays, grosbeaks, and many other birds alight in the trees' crowns to feed. Bears pull down branches and gobble leaves, twigs, and fruit altogether. Cherries that fall on the ground are picked up by rodents and ground-foraging birds such as turkeys and grouse.

Pin cherry trees line the roads veining our forested acres. They are the remnants of a much younger forest that temporarily claimed the site following heavy logging in the 1980s. Pin cherries need abundant sunlight, which is why they remain along the edges of our roads; the ones in the woods died long ago. They are short-lived trees, and many of them are now dying. I remove more fallen pin cherries from our woods road system than any other type of tree.

The droppings of wild creatures spread pin cherry seeds far and wide. (I don't know about turkeys; maybe their muscular gizzards grind up the seeds and destroy them.) The seeds are miraculous packets of life. Scientists believe they can stay viable for 50 to as many as 150 years, lying buried in the thin skin of humus that covers the forest floor. As the years pass, the seeds' tough outer coats gradually become more permeable to water and oxygen.

Then there might be a fire, a windstorm, or chainsaws and skidders to suddenly topple the mature oaks, ashes, birches, or maples whose crowns cast shade on the ground. The sun warms the ground, soil temperatures rise, and the pin cherry seeds detect the change and respond.

Exposed to full sunlight, pin cherry seedlings grow quickly. Pin cherry trees play an important role in forest development: They stabilize newly exposed soil and provide cover and light shade for the seedlings of other, longer-lived species destined to become the next generation of forest trees. In time, those trees rise above the pin cherries, and the cherries die in the shade.

But they still survive—as seeds, if not as individual trees. I find it thought-provoking and reassuring to ponder on seeds that may have fallen to earth soon after the Civil War; seeds that bide their time, carrying within their husks the germ of a new forest.

Read in depth about the types of habitat that exist (or that you suspect may exist) on your land as well as the different species of wildlife that commonly use such habitats. Learn about the animals in your state and region, including their feeding habits, cover needs, and home range sizes. Then, if there are certain kinds of wildlife whose presence you can realistically hope to encourage or attract—be they box turtles, bobolinks, bobwhite quail, or black bears—start thinking about how to make your land more hospitable to them. Keep in mind that a property that supports a variety of native plants, shrubs, and trees will tend to attract and sustain a greater number of wild animals, as well as a more diverse spectrum of wildlife.

Once you have learned which animals may inhabit your area, start keeping a list of the wildlife you see on your land, including insects, amphibians, reptiles, birds, and mammals. Take note of the plants, too. This can be an enjoyable project, especially if you don't set a deadline for completing your list or worry too much about developing an exhaustive inventory.

Head out at different times of the day, perhaps emphasizing dawn and dusk, which are times when wildlife tend to be most active. Venture afield in all four seasons. As well as wearing proper clothing, take along a field pack with water, snacks, raingear, and a notebook and writing implements. If you live in an area where ticks or biting insects are a problem, take precautions against those pests. Bring binoculars and perhaps a camera to record any finds, such as plants and insects, for later identification.

Don't just stay on the roads and trails you normally follow. Instead, mentally divide your tract into different units. On each unit, quarter back and forth systematically until you have covered your entire property. Stop in interesting and likely looking places. (You'll develop a feel for wildlife hotspots the more time you spend outdoors.) Sit down on a log or a stump

Take along your binoculars and look for animals that are using your land. Take notes on insects and vegetation, including low plants, shrubs, and trees. You can make your property more hospitable to the wildlife already present—and attract other animals that currently may be passing it by.

(or a portable hiker's chair) and let the woods or meadow or beaver pond come alive. Sit still, look, and listen. Observe for 15 minutes to half an hour or longer. You'll probably be pleasantly surprised at the number and variety of animals you see and hear.

The next step is to create a map. A simple approach is to make a sketch of your property, using a large enough sheet of paper that you can add important details. Or photocopy and enlarge the part of a U.S. Geological Survey topographical map that includes your land. You can also use Google Earth; download the free program, then enter your address in the search box. Zoom in and out to see your land and surroundings, come up with an image to create a base map, and save the image to your computer. You can also ask the USDA Natural Resources Conservation Service (NRCS) how to obtain aerial photographs covering your land. (The NRCS supports a website for each state with contact information for your local office.) Another potential source of aerial photos is your county forester or wildlife or forestry extension agent.

Are you tech savvy? Online courses and books can teach you how to use geographic information system (GIS) open-source software and freely available spatial data to create a detailed, modifiable computer-generated map of your land showing terrain and special habitat features. The web-based company TerraServer sells aerial photos as well.

Want to save money on your real estate taxes? Look into whether you can enroll your land in a state-subsidized program in which your property will be taxed for its current use—such as farming, forestry, recreation, or wildlife conservation—rather than its potential for subdivision or development. Typical names for such tax-abatement programs are Clean and Green (Pennsylvania), Use Value Appraisal (Vermont), and Present-Use Value Program for Forestry (North Carolina). To participate in such a program, you will need to own a minimum number of acres, which will vary from state to state. You may also need to hire a natural resource professional or a consulting forester to develop a land management plan or a forest stewardship plan, either of which will probably include an accurate map. Such plans may cost $500 to $1,000, although if you sign up for NRCS's Environmental Quality Incentives Program, it may cover all or part of those expenses. In any case, the money you save on taxes will probably amortize any planning or enrollment costs within a year or two.

However you develop a map, make sure it includes accurate boundaries for your parcel. Then, if they are not already represented, add special features such as forest openings, streams, spring seeps, wetlands, grass fields, shrub fields, rock ledges, logging roads, hiking trails, special trees and places, and viewsheds.

LOOK OUTWARD

When considering how best to help wildlife, take a look outward. Try to find out, at least in general,

New England Cottontail Working Landscape

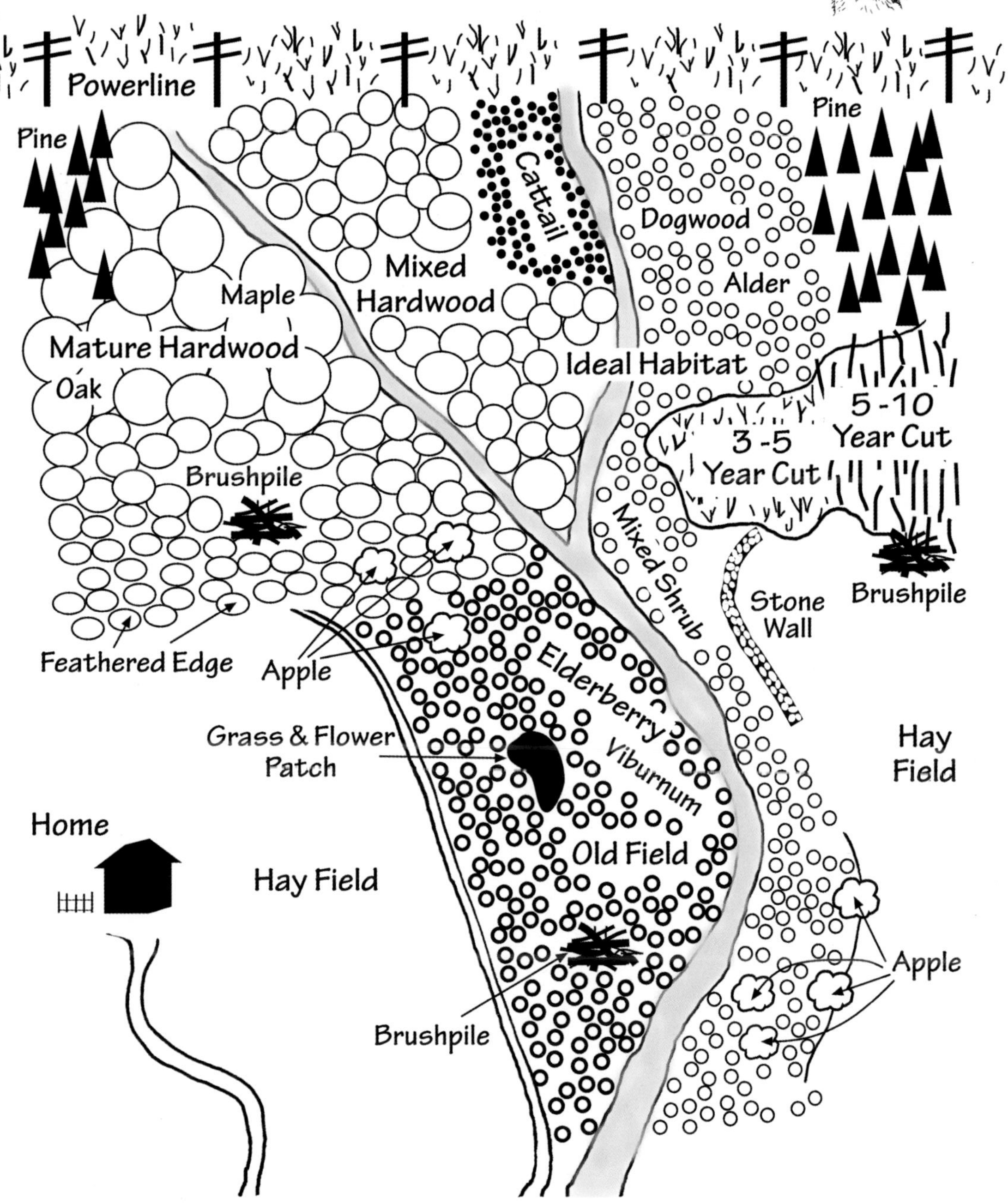

This sample map shows how a property could be managed to help New England cottontail rabbits. Something similar could be drawn up to emphasize a different kind of wildlife or for wildlife in general. It shows important habitat features such as stone walls, brush piles, and a power line acting as a travel corridor. *Paul Fusco, CT DEEP*

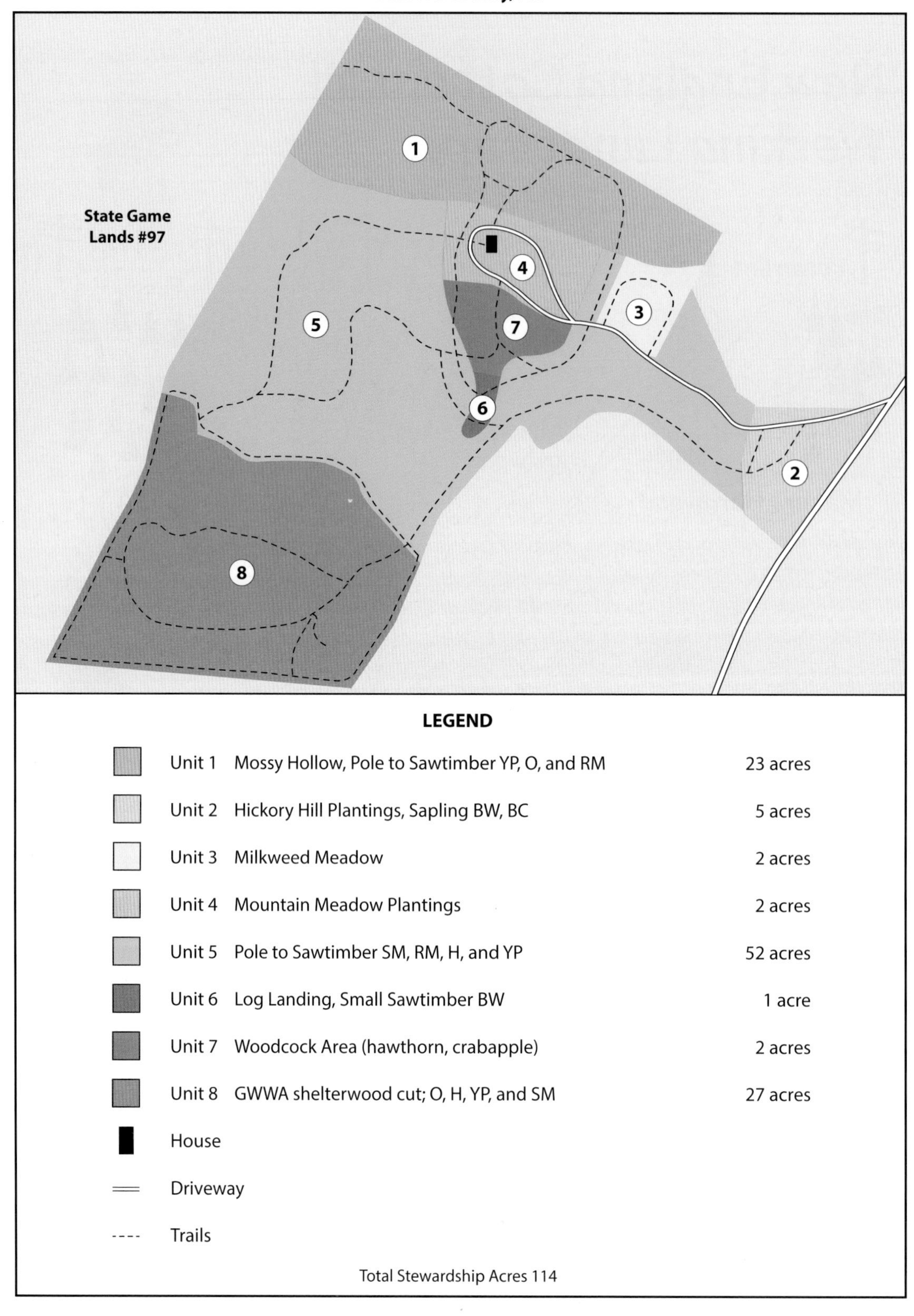

A larger-scale map depicts eight habitat management units covering 114 acres on a property on the side of a mountain in south-central Pennsylvania. The initials stand for different kinds of trees: YP for yellow poplar, BW for black walnut, BC for black cherry, SM for sugar maple, RM for red maple, O for oak, and H for hickory. GWWA stands for golden-winged warbler.
Laura and Mike Jackson

What kinds of habitat exist on neighboring properties? Evaluate an area 10 times the size of your land. Rather than duplicating a type of habitat that's already abundant, you may want to create habitat that is less common. *Ben Longstaff, IAN Image Library, http://ian.umces.edu/imagelibrary/*

what sorts of habitat exist on other lands around your property. One rule of thumb is to evaluate an area 10 times the size of your own land. This is less pertinent when you live in a town or a development where the lots are small. However, if you own five acres, look at the surrounding 50 acres. If you own 10 acres, evaluate the surrounding 100 acres. If you're fortunate enough to own 100 acres, consider the habitat on the surrounding 1,000 acres. Looking beyond your boundaries may lead you to create a type of habitat that is rare in your area. For example, I decided to keep two meadows of two acres apiece on our 120-acre Vermont farm functioning as old-field habitats, since very little similar cover exists nearby.

You are now at a stage where you can begin considering options for creating habitat. The size of your property will affect what and how much you can do to help wildlife—as well as the management approaches, tools, and processes you will end up using and the resources available to you as you plan and carry out projects.

The "order of magnitude" concept is a useful way to look at parcel size: up to one acre; one to 10 acres; 10 to 100 acres; and 100 to 1,000 acres.

John and Catherine Smith live on a 68-acre old farm in central Pennsylvania where, over the last 15 years, they have spent considerable time and money creating a wide range of habitats, as well as a place where landowners can attend workshops and seminars on how to improve their properties for wildlife. The Smiths' essay, "Managing Land Ecologically Using Perspective of Scale," employs the order of magnitude concept to examine what can be done on different-sized landholdings. (It can be found on the website of their Chicory Lane Farm, which also describes habitat projects that the Smiths have undertaken to increase plant and animal diversity on their land.) Their thoughts and experiences inform much of what follows in the rest of this chapter.

UP TO ONE ACRE

A one-acre lot, if square, would measure 208 feet on a side. Many building lots are smaller than that, down to a quarter acre (common in many subdivisions), which would be 100 feet on a side if square. If you own an acre or less, your options for creating habitat are limited by your parcel's size—but they are still very real and include choices that can help wildlife and the environment, let you learn about nature, and increase your chances of seeing birds and other animals. On a quarter-acre lot, you can put in a butterfly garden with nectar-producing flowers for adult butterflies to feed on, plus host plants on which the adults can lay eggs and, after the eggs hatch, young caterpillars can feed and grow. You can remove non-native ornamental shrubs and replace them with equally beautiful native varieties whose foliage supports insects (especially caterpillars) that birds feed on and that produce fruits that sustain both resident and migratory birds.

On an acre or less, you may not be able to conduct multiple projects—installing a small pond or wetland, planting a copse of trees, and creating a meadow—but you can probably do one of those things. Don't just wade right in; take your time. Even on a small lot, try to identify basic features and site conditions, including soil type and moisture, whether any water is present, any slope to the land and the direction it faces, the amount of light the tract receives, and plants and animals already on the site. Be sure to check with your town, township, village, or neighborhood owners' association for land-use regulations, lawn-care ordinances, or covenants that may affect what you can do.

The following are essentially landscaping projects that you can accomplish with your own labor plus some simple tools and equipment.

Pocket wetland: Look for a low, damp area that naturally holds water and add native plants around its edges. You can also create a small wetland by digging a shallow depression, lining it with a heavy-duty pond liner, adding a layer of soil to protect the liner, filling the depression with water, and planting native aquatic and wetland vegetation. A small wetland will attract birds and insects, and, depending on your surroundings, it may also draw in frogs, toads, and salamanders.

Meadow: Replace part of your lawn with low-growing native plants that provide food and cover for wildlife, plants that will need less water, fertilizer, and maintenance than groomed grass. The National Wildlife Federation estimates that the typical lawn costs hundreds of dollars per acre per year to maintain, while a meadow of perennial wildflowers can be kept up for an annual $30 per acre. Combine different species of grasses and wildflowers, depending on what sort of wildlife you want to attract. Various companies offer seed mixtures for different regions, conditions, and soil types.

Woods: Plant some native hardwood trees. You can probably fit eight to 10 trees on a one-acre lot. Oaks, maples, hickories, and cherries supply nuts, seeds, or fruit for wildlife, plus support large numbers of insects that birds feed on. The trees will sequester carbon and provide cooling shade. University extension services and state wildlife and forestry agencies are good sources of information about which tree types will do best in your area. Buy small trees or seedlings from a commercial nursery, your local or state conservation district, or through the mail. It will take years for the trees to grow tall. In the meantime, fill in between them with native shrubs and smaller mid-canopy trees such as viburnums, spicebush, elderberry, dogwoods, Juneberry, redbud, black gum, and hornbeam.

Remember, a lot of small habitat projects can add up. If your neighbors like what you're doing, they may follow your lead. Imagine what could happen if a majority of the small-lot owners in our region replaced their lawns with natural habitats. Collectively we could build a huge refuge that would attract and support a variety of wildlife while creating opportunities for individuals and families to watch and learn about animals and nature.

ONE TO 10 ACRES

Many properties in suburban and rural areas fall within this size range; in the eastern United States, about two-thirds of private woodland owners control 10 or fewer acres. If square, a 10-acre parcel would be 660 feet on a side, enclosing 435,600 square feet, an area equivalent to about seven and a half football fields including the end zones. John and Catherine Smith note that such properties are "an ideal size to learn and practice ecological principles, as well as to enjoy their benefits, without getting overwhelmed." They're also of a size that can offer significant benefits to wildlife.

Around your house, you can undertake any of the projects described in the "Up to One Acre" section. Since you have more land, you could possibly do several such projects. However, bear in mind that once you create a habitat feature, it will be hard to remove; it will be right there in front of you for years to come and could even end up being an eyesore if you lose interest and stop working to maintain it. So take your time before launching into any habitat work. Carefully and patiently explore and characterize your land. As the Smiths write, "get to know what you like to look at and where you like to be," including during different seasons. They add: "Let ideas emerge and . . . gradually coalesce into goals and plans."

Look for things that might be considered problem spots if you just wanted to have a lawn: boggy areas, places where bedrock emerges from the soil, dry knolls, steep slopes, and weedy or shrubby openings in wooded areas. These features represent potential sites for wetlands, special plantings for pollinating insects, and shrub habitats used by a variety of wild animals.

If a stream cuts across your land, you can stabilize its banks and improve food and cover resources by planting native shrubs and trees, creating what's known as a *riparian buffer*. You can restore, install, upgrade, or expand a wetland. Create grasslands and patches of wildflowers for butterflies and native bees and other pollinating insects, and to attract larger animals that use sunny, open habitats. Maintain an old field overgrown with grasses, sedges, forbs (flowering herbaceous plants), and shrubs. After identifying plants growing on your acres, you may decide that the best course of action is simply to wage war: Gradually get rid of invasive alien plants by uprooting them if they're small and not too numerous, repeatedly cutting them back so they never get the chance to produce fruit and seeds and finally die from the stress of repeated mowing, or treating them with herbicides—and planting select native shrubs in their place.

You can plant trees—both hardwoods and softwoods—to create a small patch of forest. When planted in a grid with 20 feet between trees, it will take about 100 trees to establish a forest on one acre. Before you order trees from a private or state nursery, take the time to learn about the forest communities common in the area where you live.

Your state forestry agency or university extension service can help. Let's say you own a 10-acre

Buffers of grasses, shrubs, and trees slow the flow of surface water into streams, trapping soil particles and agricultural chemicals and improving water quality. Such plantings also provide food and cover for wildlife.

tract in the Upper Piedmont region of northern Georgia. A good information source will be the website of the University of Georgia Cooperative Extension Service. There, searching on "native plants" provides a link to *Native Plants for Georgia, Part 1: Trees, Shrubs and Woody Vines*, written by a team of horticulturists, naturalists, and master gardeners. The free downloadable PDF includes a map showing the plant hardiness or growing zones for Georgia. A chapter on "Medium to Large Trees" specifies the regions, hardiness zones, and soil conditions in which different species should thrive. You may conclude that a mix of maples, oaks, and hickories would make an excellent forest habitat for your relatively dry upland site, offering important food to birds and mammals; you might decide to add in a grove of shortleaf pine and Virginia pine to provide thermal cover during cold and rainy weather.

In recent years, a formal effort has sprung up to teach owners of one- to 10-acre parcels how to develop them into natural habitats offering many more benefits to wildlife and the environment than conventional lawns and landscaping. This forest-oriented movement, designed by university extension services and scientists in the Mid-Atlantic and Northeast regions, is called *The Woods in Your Backyard*. The overall approach includes internet seminars and a 108-page illustrated workbook. The book's authors write: "One of the most effective ways to improve water quality and air quality, wildlife habitat, and natural area health is to shift areas of lawn into unmowed natural areas or woodlands. Over a period of years, this will reduce the time you spend mowing lawn and allow more time to enjoy your family, property, and hobbies."

The workbook leads you through the entire process, step-by-step. It's basically a minicourse in creating habitat for wildlife. It teaches you how to examine your land and evaluate its current value as habitat, identify potential management units, draw up maps, and develop reasonable habitat goals. It explains how to carry out practices such as making brush piles, food plots, and openings in the woods; suppressing invasive plants; encouraging or planting native trees and shrubs; conducting timber stand improvement activities; and designing and creating woods roads and trails. It even suggests a planned activity called "Reality Check: Is Your Family With You?" including a "Family Goals Assessment" that helps a family make a group decision on whether to commit the time and money to create habitat. To learn more about this program, including on-site field events, visit the websites of Penn State University Extension, University of Maryland Extension, Virginia Tech Extension, or Cornell Cooperative Extension. If you don't live in the Mid-Atlantic or the Northeast, an internet search may turn up similar programs and opportunities in your own state.

Even though *The Woods in Your Backyard* focuses on the Mid-Atlantic and Northeast, its concepts and information apply throughout the East. Another excellent aspect of this approach is that it can be scaled up to work on larger properties. On a larger holding, it can focus attention and efforts on the one to 10 acres right around your house, the part of your property where you will spend most of your time and where you'll likely see wildlife. You can apply its principles anywhere on your land.

10 TO 100 ACRES

First, some statistics—these are from Pennsylvania, but it's highly likely that similar numbers apply to other eastern states. In Pennsylvania, owners of wooded properties in the one- to 10-acre size range make up 60 percent of private forestland owners and own 7.5 percent of forested land. In contrast, owners of woodlands in the 10- to 100-acre category make up 35 percent of all private forestland owners and control a whopping 43.5 percent of forested land in the

The Woods in Your Backyard

Learning to Create and Enhance Natural Areas around Your Home

This 108-page guide promotes the stewardship of small land parcels for their owners' enjoyment and for improved environmental quality. It's written for both educators and owners of one to 10 acres of wooded or unmowed natural areas in eastern North America and for landowners who want to turn mowed areas into woodlands. The guide features more than 100 color photos, an overview of native and invasive plants, lists of activities, case studies, appendices, and a glossary.

By reading the guide, completing the activities, and reviewing the resources, you can achieve a better understanding of your land and develop a realistic strategy to achieve your goals. Topic areas discussed include:

- Understanding the benefits of managing your land;
- Mapping your property, assessing why you own the land, and deciding what you hope it will become;
- Understanding how your land functions in the larger landscape;
- Identifying habitat units on your property;
- Learning the basics of tree identification, forestry, and habitat management techniques;
- Assessing and improving your property's water resources, recreational possibilities, and aesthetic appeal;
- Choosing one or more projects to help meet your goals, setting a timetable, and recording your progress.

The Woods in Your Backyard (second edition, 2015) was written by Jonathan Kays, Adam K. Downing, Jim Finley, Andrew A. Kling, Craig Highfield, Nevin Dawson, and Joy Drohan, with a forward by Douglas Tallamy.

Keystone State. The take-home message? If folks who own 10 to 100 acres become inspired to create, refresh, and manage habitat, wildlife will receive significant benefits.

The upper end of this size category amounts to a considerable chunk of land: a square 100-acre tract would measure slightly longer than 2,000 feet, or four-tenths of a mile, per side. Among properties in this general size range are old farms no longer in agriculture; forested tracts where timber harvests took place on a large parcel, which was then broken up into smaller tracts and sold; and lands of marginal financial value sold off by large landowners.

It takes a lot of time and effort to really get to know a property of this size—to visit every nook and cranny, to identify different habitats, forest stands, and special habitat features, and to get a sense of the plants and animals they host. John and Catherine Smith suggest that landowners buy a utility task vehicle, or UTV (theirs is a John Deere Work Series Gator 6x4), to make exploring easier. Boots on the ground work, too, and provide healthy exercise in the bargain.

If you own land in the one- to 10-acre range, your property probably will not be large enough to attract the attention and services of state and federal agencies and nongovernmental wildlife organizations whose missions include helping landowners make habitat. However, if your property has more acres (25 acres may be the tipping point), a number of government agencies and wildlife organizations will be interested in offering you advice and, in some cases, financial help.

The U.S. Department of Agriculture's Natural Resources Conservation Service, or NRCS, is attuned to helping folks who own farms and forests, both working properties and vacation and residential ones. Their habitat improvement programs are funded mainly through the Farm Bill, a piece of federal legislation that comes up for Congressional review and renewal every five years. I received advice and financial aid from the NRCS for four different habitat projects here on the Butternut Farm and found their employees to be friendly, knowledgeable, and helpful.

Here's how it works. You contact NRCS, which has offices throughout the country. (I was referred to them by the consulting forester who drew up a management plan necessitated when we placed our 120 acres in Vermont's Current Use tax abatement program.) An NRCS biologist or forester will come to your land, walk it with you, and discuss ways in which you can create or improve habitat. At your request, NRCS will draw up a conservation plan, a technical document including maps and descriptions of soils, forest stands, and habitat areas, along with recommendations for projects to improve or expand habitat. Depending on the availability of funding and the characteristics and potentialities of your land, you may qualify for financial help to carry out one or more projects through various NRCS landowner-assistance programs, which currently include the Environmental Quality Incentives Program, Conservation Stewardship Program, Regional Conservation Partnership Program, Wetlands Reserve Program, and Working Lands for Wildlife Program.

As of this writing, NRCS offers advice and aid for wildlife-oriented "conservation practices," including Brush Management, Conservation Cover, Constructed Wetland, Early Successional Habitat Development and Management, Field Border, Forest Stand Improvement, Hedgerow Planting, Prescribed Burning, Riparian Forest Buffer, Stream Habitat Improvement and Management, Structures for Wildlife, Tree and Shrub Establishment, Upland Wildlife Habitat Management, and Wetland Wildlife Habitat Management. If you decide to proceed, your NRCS contact will help you fill out forms to apply for grants. You'll still need to pay out-of-pocket costs (or contribute your own materials and labor), but the funding can make habitat improvement actions affordable.

Many projects that grow out of NRCS conservation practices can be sizable and technically complex, such as creating or restoring wetlands, converting an old pasture to a warm-season grassland, or reforesting a field. Many call for large and expensive machines, or items like posts and fencing. In short, they can be more complex than what a single person, plus family and friends, may wish to undertake—which means you will need to hire outside contractors. NRCS keeps lists of contractors who can do such work. Remember, too, that when you accept funds from NRCS, you will have to fill out fairly complicated forms (your NRCS field person will help you with that), and you will be signing a legal contract that you must honor—something that can be fairly stressful to many people. Agreeing to and carrying out a large project, or a suite of smaller projects, is a significant step that calls for honest and logical forethought and planning before you make what may be a several-year commitment. You may decide to start off with a small project and see how it goes before committing to a larger endeavor.

Your NRCS contact may advise you that your land qualifies for the Conservation Reserve Enhancement Program (CREP), administered by the USDA's Farm Service Agency. In CREP, poor-quality farmland is taken out of agriculture and, following habitat enhancement, is placed in conservation reserve status under a voluntary lease lasting 10 or 15 years. After receiving a cost-share of up to 50 percent of the expenses for any habitat improvements, you will be paid an annual rental fee for however many acres are set aside for conservation purposes. Farmers often take advantage of this program, as do private landowners.

A second federal agency to contact is the U.S. Fish and Wildlife Service (FWS). Through their Partners for Fish and Wildlife program, they offer a broad range of services and are creative and flexible in how they approach helping you help local wildlife. For any project that affects or involves a body of water—whether a stream, pond, or wetlands—you need to involve your local Fish and Wildlife Service office. Service habitat biologists can help design projects to enhance or create wetlands, or improve streams and help pay for those improvements, or put you in touch with other conservation entities, such as state agencies or nongovernmental wildlife organizations, which can also help with funding. FWS employees can help you obtain permits, keep watch over contractors' work, and make sure the completed project functions the way it's supposed to. They help with streambank restoration, planting shrubs and trees to create streamside buffers, and controlling or eradicating invasive plants. Sometimes they work in concert with the NRCS, state environmental agencies, and nonprofit wildlife organizations.

Some state conservation agencies have private landowner programs; they will send a forester or biologist to see whether your land qualifies for funding to make habitat. Many of the projects directly benefit wildlife, and others are aimed at improving forest health, accessibility, or potential for future timber sales. For example, I received advice from our county forester, with the Vermont Department of Forests, Parks and Recreation, on how to improve the woods road system on our land. He arranged for a small grant that paid for some of my time removing blowdowns and opening up unusable roads, plus funds to hire a local contractor with a small excavator who fixed sections of our roads that had been cut up by earlier illegal all-terrain vehicle use, then dug water bars to shunt water off the roads and prevent future erosion.

Among the nonprofit wildlife organizations that actively help landowners make wildlife habitat are the Audubon Society, Ruffed Grouse Society, National Wild Turkey Federation, Pheasants Forever, Quail Forever, and Ducks Unlimited. Except for Audubon, which has a general emphasis on birds, all of these organizations focus on creating habitat for birds that are hunted. However, habitat made for game birds also benefits a wide range of other wildlife that is not hunted.

A final thought to keep in mind: You can do a whole lot to improve a grassland, shrubland, wetland, or forest without needing assistance, oversight, or financial aid. You just have to learn what's present on your land, come to a logical conclusion about what you want and are able to accomplish, determine efficient and effective ways to reach your habitat goals, and then invest enough time to achieve them. Making habitat with a brush hog or a chainsaw can be among the most satisfying and enjoyable pastimes you will ever undertake.

100 TO 1,000 ACRES

A thousand acres takes in about one and a half square miles; if square, it would measure a shade less than one and a quarter miles on each side. Within this category, parcels of 100 to several hundred acres may be owned by individuals, couples, or families. Tracts that approach a thousand acres may be owned by either related or non-related people—for example, a property that has been in the same family for generations, or a hunting camp owned or leased by dozens of members.

Parcels that are 100 to several hundred acres may include both open and forested land, probably with more forested than open acres. Larger tracts, especially those approaching 1,000 acres, tend to be mostly or completely forested. Folks who own 100 acres may not expect their land to generate significant income—like my wife and me, with our 120-acre Vermont farm. However, those with larger properties may need the land and its resources to generate income and help pay property taxes or the upkeep of one or more residences or a hunting camp or lodge, and harvesting trees can help bring in that income. Planned wisely and carried out carefully, logging can improve the quality of the woods (at least from a commercial standpoint) by removing unhealthy or crooked trees. It can also be used to promote diverse tree ages and species, something that helps boost numbers and kinds of wildlife. Sustainable forestry practices can provide our society with useful products while maintaining or increasing the value of standing timber.

A good first step is to engage a professional consulting forester to help you develop a management plan spanning years and decades and suited to the types of trees on your land whether they are slow-growing oaks or quick-growing aspens. A good forester will help you decide how to deal with invasive shrubs, a large deer population, or tree diseases and insect pests.

Consider enrolling in a forest stewardship program, administered nationally by the USDA Forest Service and statewide by different natural resource agencies. Such programs generally offer classroom and field training in subjects such as biodiversity, forest ecology, silviculture (practices contributing to the development and care of trees and forests), wildlife science, and environmental resource management. In Pennsylvania, for example, the Forest Stewards Program is sponsored by four entities: the Penn State University Center for Private Forests, the Pennsylvania Department of Conservation of Natural Resources, Penn State University Cooperative Extension, and the USDA Forest Service.

As mentioned in the preceding section focused on properties of 10 to 100 acres, you can also ask for help from the USDA Natural Resources Conservation Service, or NRCS. If you have a project in mind that involves water resources—revegetating a streambank, building a vernal pool, restoring or expanding a wetland, improving a trout stream, and so on—get in touch with your local or state office of the U.S. Fish and Wildlife Service.

4

Forests for Wildlife

EASTERN NORTH AMERICA IS A GREAT PLACE FOR GROWING TREES: well-watered, generally fertile, and with an ample growing season. Trees in the region include *coniferous* types (also known as softwood, cone-bearing, or evergreen trees), such as pines, spruces, and firs; and *deciduous* species (hardwood or broad-leaved trees), including maples, ashes, beeches, birches, and oaks. Most tree species in the East are deciduous, shedding their leaves each autumn. When mature, some kinds are large and tall, their uppermost branches forming the forest's canopy, while others are smaller and shorter. Some need full sunlight; others can grow in partial or full shade. Beneath the trees may be shrubs, vines, and low plants. All of these components function together to create forest communities and ecosystems used by wildlife.

Forests yield sustainable wood products, sequester carbon, and provide wildlife with food and cover. Forest types vary throughout the East. This Vermont woodland has evergreen trees mixed with hardwoods—mainly red and sugar maples—displaying brilliant autumn foliage.
Tom Berriman

Douglas Tallamy is an entomologist and wildlife ecologist with the University of Delaware and the author of two influential books about protecting and using native plants to improve the environment and support wildlife. In a foreword to the useful book *The Woods in Your Backyard*, he writes: "Quite simply, it is woodlands that produce most of the living things that constitute nature in the eastern United States…it is [this] biodiversity that produces the oxygen we breathe, cleans and stores our fresh water, sequesters carbon, builds and stabilizes topsoil, moderates severe weather, and protects our watersheds."

A healthy forest is a vibrant, fascinating place. Forests draw people in search of natural beauty and solitude, and seeing wildlife greatly enhances that experience. Others venture into the woods to watch birds or to hunt; still others appreciate the economic returns that a working forest can deliver. Many of us have achieved the goal of owning a piece of forested land. Today, millions of private landowners make important decisions that affect their forests' health as well as the wildlife that can be found there.

FOREST HISTORY

Centuries ago, mature forests covered most of the East. Openings in that heavily wooded landscape included drainages where beavers had dammed streams, creating standing water that killed trees; after the beavers exhausted local food sources and moved on, their dams were breached, the ponds drained, and grasses, sedges, shrubs, and young trees grew in the resulting "beaver meadows," which were extensive in many areas. Lightning caused fires that burned forests and set them back to earlier growth stages. Ice storms, floods, and hurricanes also created gaps in woodlands. Native Americans cut trees for firewood and to make clearings for their villages and fields for growing crops. They also periodically set fires to bring on a flush of low, nutritious vegetation that drew in the animals they hunted, including elk, bison, deer, rabbits, turkeys, and grouse. Repeated fires resulted in persistent meadows, savannas, and other openings. Fires also favored trees with special adaptations that protected them from the effects of fire, such as oaks and certain kinds of pines.

To European settlers, the forest seemed to offer an inexhaustible supply of wood for building and heating houses and making various products, from pencils to clipper ships and cradles to coffins. By the mid-seventeenth century, colonists had begun clearing the forest to make farmland. In central Massachusetts, virtually all of the original woods were gone by the early 1800s, with thousands of small farms in their place. Massachusetts, Connecticut, and Rhode Island were 70 to 80 percent open. In the decades leading up to the Civil War, people in the Southern Piedmont cleared off much of the forest and replaced it with fields for growing cotton and other crops. Decades of widespread rapacious logging led to timber shortages in the Northeast; large-scale cutting shifted from Maine and New York to Pennsylvania, and then, in the late 1800s, it shifted west to the Great Lakes states. Some logged-off tracts grew back as forest, but in general, the percentage of forested land remained low as small-scale agriculture dominated.

In the 1900s, people began abandoning marginally productive farms in New England and the Mid-Atlantic and relocated to more fertile land in the Midwest. Others gave up farming altogether and moved to cities and towns. Shrubs and saplings reclaimed the abandoned farms, which gradually became forest again. However, it was a new forest, since chestnut blight had nearly eliminated the American chestnut (formerly the most abundant tree in the East) and sun-loving trees such as white pine, black cherry, red maple, and various oaks thrived.

Today, New York is 63 percent forested. Pennsylvania is 55 percent forested, with the state growing more wood than it consumes every year. Maine is 90 percent forested; Vermont is about

75 percent wooded. In the Upper Midwest, 53 percent of Michigan, 46 percent of Wisconsin, and 34 percent of Minnesota are covered in forest, with the northern regions of those states more heavily forested than the southern agricultural zones. Forests cover five million acres in Indiana, or about 21 percent of the state. According to the USDA Forest Service, Georgia has the greatest forest coverage of any Southern state, with 24.8 million wooded acres making up 67 percent of the land cover. Even heavily developed New Jersey has approximately two million forested acres out of a total of five and a half million acres, making the Garden State about 35 percent wooded. Most of the region's forests are middle-aged, currently around 80 years old. Such woodlands often lack what foresters term "age-class diversity," with only a small percentage of young forest (less than 10 years old) and old forest (100 years and older) present.

Across the region, the trend is toward smaller forested properties and more individual ownerships of forested land, as large tracts are subdivided and subdivided again. According to *My Healthy Woods: A Handbook for Family Woodland Owners Managing Woods in New Jersey*, published by the Aldo Leopold Foundation and the American Forest Foundation, "The average length of land ownership has dropped to less than 10 years" in that state. "More parcels and higher turnover means many new landowners are on the landscape." A similar situation exists in other states, affecting timber management and harvesting options, the spread of exotic invasive plants, and the quality of habitat available to wildlife.

NAMING THE WOODS

Ecologists have developed several systems for classifying the forests of eastern North America. Richard Yahner, in *Eastern Deciduous Forest: Ecology and Wildlife Conservation*, recognizes four general regions: The *Northern Hardwoods Forest* spans southern Canada and northern New England, New York, and Pennsylvania and extends to the Upper Midwest; beech, birches, maples, and northern red oak are prominent, along with pines, spruces, and firs. The *Central Broad-Leaved Forest* reaches from southern and coastal New England south and west to the Great Plains and along the Appalachians; oak, hickory, ash, basswood, buckeye, tulip tree, red maple, hemlock, and white pine are common. The *Southern Oak-Pine Forest* exists from southern New Jersey south to Florida on the coastal plain and the Piedmont, and west to the Mississippi River, with six species of pines and eight species of oaks, plus other conifers and hardwoods. The *Bottomland Hardwood Forest*, centered on the deep alluvial soils of the Mississippi River drainage in Louisiana, Mississippi, Arkansas, and neighboring states, has maples, oaks, gums, cottonwoods, bald cypress, river birch, and other trees.

The USDA Forest Service breaks things down further, identifying 11 different ecoregions in the East supporting different forest types or "provinces." Here in northern Vermont, the woods on our farm are part of the "Adirondack—New England Mixed Forest." In central Pennsylvania, where we used to live, our 30 wooded acres were in the "Central Appalachian Broadleaf Forest." A wooded area in northern Florida would be in the "Outer Coastal Plain Mixed Forest." Forest stands in the central Midwest would be in the "Prairie Parkland (Temperate) Province."

You can determine which type of forest prevails in your own area by searching the internet for "forest regions of North America." Within a forest type, the species and size of trees growing on a site can vary, depending on local seed sources; soil quality, depth, and moisture; elevation; the angle of any slope and the direction it faces; natural disturbances; and humans' past and current land-use practices.

FOREST SUCCESSION

Forest succession is the process by which a forest regenerates, or grows back, following a natural

Stands of young trees offer excellent hiding cover combined with plentiful food, including insects, buds, leaves, fruits, and seeds. In many parts of the East, young forest is a rare habitat for wildlife.
James Oehler

disturbance such as a flood, fire, ice storm, or windstorm, or a human-caused disturbance such as logging or the abandonment of an old farm field. The presence and abundance of different wild animals change during the successional process, as the habitat provided by the regrowing forest gradually becomes more suitable for some kinds of wildlife and less suitable for others.

Ecologists identify the following stages:

Young Forest: Trees are zero to 10 years old (they can also be as old as 15 to 20 or more years, depending on local site and tree-growing conditions). Other names are *successional forest* and *early successional forest.* After a disturbance or farm abandonment, seeds and nuts of trees send down roots and begin to grow. Some types of trees can send up shoots from their root systems or from the base of the trunk if the tree has been injured or cut down. In a young forest, grasses and forbs carpet the ground, and shrubs may mix with the small trees. Often, many of the trees are "pioneer species," quick-growing types that need full sunlight. The leaves they drop each year help build up humus, enrich the soil, and increase soil moisture. Dense young forests provide excellent hiding and resting habitat for wildlife, as well as food in the form of abundant insects and fruits. Most ecologists agree that there is currently a shortage of young forest habitat in the East; conservation agencies have begun making young forest on lands they manage and encouraging private landown-

Sometimes called "pole stage" woods, this forest growth stage generally has less food and cover for wildlife than younger or older woodlands.

ers to do the same. Young forest can be created through timber harvesting, letting trees take over an old field, planting trees, and conducting prescribed burns that kill older trees and spur younger growth.

Intermediate Forest: Trees are 10 to 40 years old, with trunks four to 10 inches in diameter at breast height. They are not yet large enough to produce reliable crops of nuts, such as acorns and hickory nuts, and they lack cavities in their trunks and limbs to provide cover for birds and small animals. Their foliage is out of reach for most browsing wildlife. The trees' canopies cast shade on the ground, and fruit-producing shrubs and ground plants dwindle. Also called "pole stage," this is often the least productive forest habitat for many kinds of wildlife.

Mature Forest: Typically 50 years old and older, the trees are large enough to produce abundant nuts (often referred to as "hard mast") and fruits ("soft mast"). Dead wood and leaf litter build up on the ground, and trees begin to develop cavities in their limbs and trunks, which wildlife use for shelter, rearing young, and hibernating. Shade-tolerant trees and shrubs persist in the understory. If a natural or human-caused disturbance takes place, the habitat will revert to a younger growth stage.

Old-Growth Forest: It takes over 100 years, and as many as 200 to 300 years on some sites, for a forest to reach this venerable stage. Very little true old-growth forest exists in the East today; most of it consists of protected remnants of the earlier original forest. However, some second-growth forests have begun to approach this highly productive age. In a classic old-growth stand, trees are large, with massive trunks free of lower branches, deeply furrowed or plated bark, huge roots, and thick limbs. Shade-tolerant trees may dominate the site, with smaller trees of the same species surviving in the midstory and understory, ready to grow rapidly toward the sun if an opening appears in the forest canopy, such as that caused by a mature tree dying or falling.

Trees in mature forests are large enough to produce abundant nuts and fruits and to develop cavities in their limbs and trunks that animals use for sheltering and nesting.
© iStock.com/Lightwriter1949

WILDLIFE SKETCH

Mood Music

The singing of male wood thrushes graces mature forests throughout the East. "This song we shall remember though we forget all others."
Ed Schneider

TO MY EAR, it's the sweetest music in the woods: the springtime calling of wood thrushes. Their song goes like this: *ee-o-lay*, the notes pure and lilting, with the first note high, the second note lower, and the third note ascending again. Repeated many times. I have never tired of that beautiful melody. It reminds me of the instrument called the recorder, played by a musician trying out variations on the same theme, experimenting with slightly different tones and pitches.

The wood thrush's calling inspired Henry David Thoreau to write: "How cool and assuaging the thrush's note after the fever of the day!" and "Whenever a man hears it, he is young, and Nature is in her spring."

A wood thrush is a subtly pretty bird, about eight inches long, a tad smaller than the more well-known robin, which is a close relative. A wood thrush sports a dark-speckled pale breast and a rusty red back and crown. Wood thrushes inhabit large blocks of mature forests, where they generally stay in the understory and on the ground. They're not easy to spot, even when the males are singing their hearts out.

I moved from central Pennsylvania to northern Vermont in 2003. Here on our wooded acres, unfortunately, we have no wood thrushes (although we do have hermit thrushes, whose songs are almost as lovely). Whenever I go back home to Penn's woods, I try to put myself in places where I can hear thrush music. A recent May morning found me in a state forest in central Pennsylvania. Rain the night before had left the woods damp and quiet. Quiet save for the stream burbling over sandstone rocks and the male wood thrushes whose calls percolated in counterpoint to the water's song.

Walking a trail, I heard thrushes on my right and left. At one point, I stopped where I could hear no fewer than three birds. Notes emanated from a thicket of glossy-leaved rhododendron. From a tangle of striped maple came another, slightly different interpretation of *ee-o-lay*. A third call bubbled up from a shady nook behind me.

The thrushes suggested players in a reclusive orchestra, tuning their instruments before the maestro comes on stage. I stood and listened for a long time before moving off, and it was hard to keep in mind that all the singing—at least from the standpoint of the thrushes—probably had nothing to do with creating beautiful music and everything to do with defending a breeding territory.

And yet so mellifluous, so haunting. Arthur Allen, an early twentieth-century ornithologist, wrote in an essay about the wood thrush: "This song we shall remember though we forget all others."

There are many cavities in the trees. Standing dead trees abound; fallen logs lie decaying on the forest floor. The soil has a thick layer of humus, and soil moisture is high. Biodiversity tends to be greater in old-growth stands than in younger ones. Other names for this growth stage are *primary forest*, *climax forest*, and *late successional forest*.

VERTICAL STRUCTURE

This term refers to layers of vegetation in a woods, including small plants on the forest floor, shrubs in the understory, medium-sized trees in the midstory, and tall trees in the overstory. Today, many Eastern woodlands have only two layers of vegetation: the plants on the ground and the foliage of the overstory trees. Other layers in between—shrubs, vines, tree seedlings, saplings, and small trees—may be absent, especially if a high population of deer has repeatedly browsed off such growth.

A forest with multiple layers of vegetation provides food and cover for many kinds of wildlife. A diverse vertical structure lets birds find places to perch, forage, hide, nest, and raise young. Cerulean warblers nest in the overstory. Wood thrushes usually nest five to 15 feet above the ground; they thrive in woodlands that have shrubs or small trees at midstory height. Worm-eating warblers hide their nests among dead leaves on the forest floor, usually against the base of a shrub or a sapling, and they feed on insects in the leaf litter, on tree bark, and on foliage.

Vertical structure provides cover, travel corridors, and feeding opportunities for wildlife. Animals that spend time in trees are considered *arboreal*. Porcupines ascend trees to feed on bark, twigs, buds, and leaves. Gray squirrels move about by leaping from branch to branch. White-footed mice scamper along tree limbs. Excellent climbers, fishers hunt for porcupines and squirrels in trees. Some snakes climb into trees to hide and find food. Gray tree frogs, seldom seen but common throughout the East, rest on branches and leaves;

A forest with many layers—from logs and leaf litter on the ground all the way up to the canopy trees' topmost limbs—provides a range of different cover and feeding options for wildlife.

in summer and fall, they hunt for insects at night, their sticky toe pads letting them clamber about with agility.

EDGE AND FRAGMENTATION

The area where two different stages of forest growth meet—or where two different ecological communities merge with each other—is called an "edge." An edge can exist between a field and a stand of mature trees; where low-growing and older, taller trees merge; or where a woods road or a log landing interrupts the forest. An edge can occur between woodlands and wetlands, or woodlands and grasslands. Because two varieties of habitat intersect, a greater diversity of wildlife will often be found along an edge, compared to the two habitats by themselves.

The following is from a 1970s university extension publication: "It is to the landowner's advantage to maximize the amount of edge in a woodland." That view has changed somewhat as research into the habits and populations of various animals has shown that too much edge can cause problems for some wildlife. Raccoons, skunks, foxes, and feral cats hunt along edges and may prey heavily on wildlife there, including songbirds that nest low in trees, in shrubs, or on the ground. Brown-headed cowbirds are *nest parasites:* The females surreptitiously lay their eggs in the nests of other birds, which then raise the baby cowbirds, with fewer (if any) of their own offspring surviving. Brown-headed cowbirds feed in open areas such as meadows, pastures, lawns, and fields, and then they duck into woodland edges, gaining access to the nests of birds that breed in forests. Too much edge habitat can increase nest parasitism by cowbirds, reducing other birds' populations.

A high percentage of edge occurs where the forest has been broken up into small patches by development or farmland. *Forest Management for New York Birds: A Forester's Guide*, by S. M. Treyger and M. F. Burger (Audubon New York, 2017), recommends that in areas where the landscape is less than 70 percent forested, and where forest cover is fragmented by agriculture or development, landowners who want to help forest birds should keep large, contiguous tracts of mature forest intact. Such core forest areas are important to *area-sensitive* species (scarlet tanagers and wood thrushes, for example) that require large patches of woodland to establish breeding territories, nest, and rear their young. Area-sensitive forest birds need a minimum of 200 acres of forest.

Still, it can be good to have some gaps within an extensive forest. A small clearcut timber harvest—also called a patch cut—can help birds and other wildlife by creating a food hotspot. After trees are removed, sunlight reaches the ground and spurs the growth of low vegetation. There, adult birds can find insects to feed their nestlings. After young birds leave the nest, their parents can take the fledglings to those thick areas, where the inexperienced youngsters can learn to feed themselves and find plenty of nutritious fruits and insects while the closely spaced shoots of shrubs and small trees shield them from predators such as hawks.

Other landowners may decide to aggressively harvest timber to promote edge habitats because they want to encourage certain kinds of wildlife. A friend here in Vermont periodically logs tracts of forest on his land, both to make money and to create large patches of young forest that attract and support ruffed grouse and American woodcock—game birds that my friend, using his well-trained pointing dogs, hunts in autumn. In an old orchard, my friend has "day-lighted" dozens of apple trees by cutting down nearby taller trees, so that they won't shade out the apples, whose fruits feed grouse, foxes, coyotes, deer, and black bears. Woodcock probe for worms in the rich soil under the revitalized apple trees.

MANAGE YOUR FOREST FOR WILDLIFE

There's an old forestry saying that you can do three things with a piece of land: Plant something on it, cut something down, or do nothing. You can convert open land to forest by planting trees. You can cut down existing trees: harvest them, in foresters' parlance, with several possible goals in mind. Or you can just wait and let the trees grow. This is perhaps an over-simplification, but it is basically accurate. Let's look in greater detail at the three approaches outlined above.

Planting Trees

Planting large numbers of trees and caring for them so that they survive is known as *afforestation*. It can be complicated and expensive. In Chapter 3, I referred to the essay "Managing Land Ecologically Using Perspective of Scale," by John and Catherine Smith. The Smiths are Pennsylvania landowners with much practical experience in making wildlife habitat—including converting a 13-acre old field to a hardwood forest. The following section draws on information in the Smiths' essay, plus John's answers to some questions I asked.

The first step in establishing a forest stand is deciding which types of trees will do best on your site. Get your soil tested. (Kits can be obtained from university extension services, agricultural supply centers, and gardening stores. You put a sample of your soil in an envelope, mail it to a soil lab, and they send you a report describing the

Before planting trees—the technical term is "afforestation"—figure out which types will do best on your land. Bare-root seedlings can be purchased from private nurseries or those run by state forestry and wildlife agencies.

LANDOWNER STORY

The Music in the Pinelands

Over the last 16 years, with help from The Nature Conservancy and federal and state agencies, Bill Owen has planted some 625,000 longleaf pine seedlings on his family farm. *USFWS*

"I HAVE TO ADMIT THAT I HATED LIVING HERE WHEN I WAS GROWING UP," says Bill Owen, describing his family's farm near the village of Yale, in Sussex County, southeastern Virginia. "I wanted to be away, in the city, in a university setting." Owen got his wish, becoming a classical musician, a Fulbright scholar in Vienna, a university teacher, and a church choir director and organist.

Along the way, he inherited the farm from his parents.

He had little practical experience with managing land. "I wasn't particularly motivated to be a landowner," he says. However, he decided to hold on to most of the 700 acres he inherited, and he pondered what to do with the property. Around 200 acres were being farmed, and the rest was in timber stands of varying ages. In southern Virginia, and in the South in general, many woodlands consist of pines planted and managed as a crop. It's forestry on an industrial scale, a highly mechanized process in which trees get cut at young ages and turned into products such as paper and plywood. "I don't like what that kind of forestry has done to the landscape,"

Owen says. "I wanted something different. I knew about The Nature Conservancy, so I called them up."

Brian van Eerden, a specialist in longleaf pine ecosystems and the use of fire to restore natural habitats, visited Owen's land. They examined a one-acre logging deck where tree-harvesting equipment had compacted the sandy soil and decided to do a test planting of longleaf pine seedlings. If the trees could grow on that compromised site, van Eerden reasoned, they would likely prosper anywhere on Owen's property.

"That was in 2002," Owen says. "The pines on that acre have really grown since then." That first acre became the start of a magnificent obsession for Owen, who, with help from The Nature Conservancy and several state and federal agencies, has transformed his woodlands over the succeeding 16 years by planting more than 625,000 longleaf pine seedlings on some 1,000 acres, with another 324 acres scheduled for planting in 2018.

Longleaf pine gets its name from its long, pliant needles, eight to 18 inches in length. Before European settlement, it was the predominant forest tree on the Southeastern Coastal Plain, growing in a verdant swath from Virginia to Florida and west to eastern Texas, covering some 92 million acres—an area almost as large as California. Today, about 5 percent of that native range remains as longleaf pine woodland, considered to be one of the most highly endangered ecosystems worldwide. Most of the original forest was logged off, with the strong, straight-grained wood used for products as various as ships' masts, factory beams, and house floors. Many of the trees were exploited for their resinous gum, which was turned into pitch, tar, turpentine, and other products. Forests were also converted to cropland and taken over by development. Without mature longleaf pines on hand, not enough seeds were produced to naturally reforest logged areas. Over the years, timber companies and private landowners replanted vast acreages to loblolly pine and slash pine, both of which grow faster and yield a return on investment sooner. In Virginia, ecologists believe longleaf pine once cloaked around 1.3 million acres. Now, fewer than 200 mature native longleaf pines remain in the entire state.

In Virginia and throughout the South, both domesticated pigs turned loose to graze and burgeoning numbers of wide-ranging feral hogs eat the growing tips of longleaf pine when it's in the "grass stage," for the first few years after sprouting, which kills the immature pines. The nationwide effort to suppress wildfires also has harmed longleaf pine, which needs frequent low-intensity burns to thrive. Fire, whether caused by lightning or humans, exposes areas of bare soil where the winged seeds of longleaf pine can germinate. Without fire, longleaf pines have difficulty reproducing and are likely to be overshadowed by other, faster-growing trees.

Today, conservationists use controlled burning to help longleaf pine in key areas in the South. Periodic low-intensity fires kill vegetation that otherwise would outcompete longleaf pine, including loblolly pines and hardwood trees such as oaks, gums, and hickories. Repeated controlled fires also reduce the amount of litter on the ground—pine needles, fallen limbs, dry leaves, and other debris—that, should it build up, can spur uncontrolled wildfires that endanger both wildlife and people.

LANDOWNER STORY

Longleaf pines are beautiful throughout their lifespan, from the fountain-shaped needles of the grass stage, to the "bottlebrush" growth of five-foot-tall seedlings, to the towering mature trees that stand like columns above an understory full of native shrubs and low plants. More than 2,500 species of plants, including sedges, grasses, wildflowers, and orchids, grow beneath longleaf pines. The longleaf pine ecosystem is among the most biologically diverse of all forest types in North America. This ecosystem is home to 100 kinds of birds, 36 kinds of mammals, and 170 types of reptiles and amphibians; the U.S. Fish and Wildlife Service classifies 29 of its resident animal species as threatened or endangered.

The keystone wildlife species of longleaf pine is the red-cockaded woodpecker. This medium-sized black, white, and red woodpecker nests in older pines that are dying but are not yet dead: alive and solid on the outside, with an interior infected by red heart rot, a fungal disease that softens the wood, letting the birds use their bills to excavate cavities for nesting and sheltering. Red-cockaded woodpeckers create cavities in loblolly pines, but they enjoy greater nesting success when they make their holes in longleaf pines, which live longer and get bigger than loblollies.

Longleafs produce abundant sticky resin around the entrances to red-cockaded woodpeckers' cavities, which may deter nest predators such as squirrels and snakes. The woodpeckers find plenty of food among longleaf pines, using their bills to flake off bits of bark and expose insects. Red-cockaded woodpeckers can move easily through the open spaces beneath the crowns of mature longleaf pines as the birds fly about in search of food. And when landowners or conservation agencies create habitat for red-cockaded woodpeckers, a suite of other wildlife benefits as well.

"I love the whole longleaf ecosystem," Owen says. The wind soughing through the pines' crowns, the *bob-bob-white* calls of quail, the raucous gobbling of wild turkeys, the high-pitched *pings* and *preeps* of toads and tree frogs. Bats catch flying insects among longleaf pine. Summer tanagers and yellow-throated vireos feed in the upper and lower levels of a pine stand, while Bachman's sparrows forage in the grass. Fox squirrels, deer, butterflies, and pollinating insects are a few of the other animals that live in and under longleafs. In states south of Virginia, gopher tortoises trundle across the sandy soil of longleaf habitats. Indigo snakes, scarlet kingsnakes, and corn snakes with dazzling red, gray, and orange markings hunt for the many rodents that live in the undergrowth—as do canebrake and diamondback rattlesnakes.

The importance of longleaf pine to nature and wildlife has given rise to an agreement, made in 2010, between the federal departments of Agriculture, Defense, and the Interior to commit to restoring eight million acres of longleaf pine. Following that agreement, the USDA's Natural Resources Conservation Service established a Longleaf Pine Initiative through which NRCS and various state agencies offer advice and funding to help landowners in nine southern states profitably grow and harvest trees while protecting and restoring longleaf pine; the Initiative has a goal of an additional 4.6 million acres of longleaf pinelands by 2025.

For more than three decades, The Nature Conservancy (TNC) has been a major player in bringing longleaf back. TNC cooperates with timber companies to develop strategies that maintain and improve intact longleaf tracts; they also work with the

The longleaf pine ecosystem is among the most biologically diverse of all forest types in North America. It is home to 100 species of birds, 36 kinds of mammals, and 170 types of reptiles and amphibians.
Mark Godfrey, The Nature Conservancy

U.S. Department of Defense to protect key longleaf areas on military bases while preserving the military's ability to train troops and conduct maneuvers. Another nongovernmental organization, the Longleaf Alliance, works to ensure a sustainable future for the longleaf pine ecosystem through conservation partnerships, landowner assistance, and science-based education and outreach.

Owen has received funding from the NRCS and three Virginia state agencies to plant longleaf pine. Dave Byrd, a habitat biologist with the U.S. Fish and Wildlife Service's Partners for Fish and Wildlife Program, provided key advice and arranged funding to help Owen carry out management tasks. The Nature Conservancy has continued to aid Owen's effort since that one-acre experimental planting back in 2002, through funding and technical assistance; Owen jokingly refers to TNC's Brian van Eerden as "a Rhodes Scholar in longleaf pine."

Creating a healthy longleaf forest is neither a swift nor an easy undertaking. On Owen's land, the process generally goes like this: A tract of forest, usually dominated by loblolly pine, is logged in a clearcut, or even-aged, timber harvest. "After cutting, we let the site grow back for one summer," Owen says. "Then we spray herbicide, typically through an aerial application delivered by a helicopter. This helps get rid of trees that would compete with the young longleaf pines." A fire crew then conducts a prescribed burn, which further stresses competing plants while scorching the land and exposing mineral soil, creating a clean site for planting.

Crews of workers tramp over a planting site, using special picks to dig holes of the correct size and shape for longleaf seedlings. "You can't plant them too deep," Owen says, "or they won't survive." He buys his seedlings from a Weyerhaeuser nursery and from the North Carolina Department of Forestry, which has been growing longleaf pine of native Virginia stock for the Virginia Department of Forestry; the idea is to plant trees with the best chance of prospering under local soil and weather conditions. "As often as I can, I try to follow along behind a crew as they work," Owen says. "I know the land well enough that I can suggest they plant a few extra seedlings in this place, or avoid that spot because sometimes it gets too wet." Owen estimates it costs between $350 and $400 per acre to prepare a site and plant longleaf pine on it.

The longleaf seedlings grow for around five years, adding a little height but mostly sending down a long taproot that can reach a depth of six to 10 feet. "At that point, we do the first burn," Owen says. "When they're in the grass stage, it's okay to burn—the fire just rolls right through, burning the native grasses and underbrush. The thick tops of the small pines protect the plants' growing tips."

After young longleaf pines develop those deep roots and bank up energy in their root systems, they start shooting upward, sometimes growing several feet in only a couple of months. During this "bottlebrush phase," no horizontal branches are formed—the plant is trying to elevate its sensitive growing tip to a height where it won't be harmed by fire.

Once the pines are big enough, Owen tries to burn each stand every two years. "This is an ongoing effort," he says, "and I'm incredibly fortunate to have a good relationship with TNC."

The Nature Conservancy has deep experience in using prescribed burning to improve the health of its lands. The largest private landowner of longleaf pine stands in the country, TNC has restoration projects underway in all nine states within the pine's historic range and manages more than 156,000 acres of longleaf through outright ownership or purchased or donated easements. The organization's Piney Grove Preserve, about 20 miles from Owen's land, includes 3,200 acres of mature pine. There, TNC burns over 1,000 acres of pine forest every year, using experienced burn crews made up of TNC, federal, and state employees.

Grants from the U.S. Fish and Wildlife Service, NRCS, the Virginia Department of Forestry, and TNC have covered most of the costs of replanting Owen's timber stands to longleaf. He has used income from timber sales to buy seedlings and have them planted—and to purchase more forestland adjoining his farm, where, after harvesting the timber, he converts the woods to longleaf as well.

TNC's Piney Grove Preserve includes old-growth loblolly pines that are home to a growing population of red-cockaded woodpeckers. The Commonwealth of Virginia recently bought 4,400 acres next to the preserve to create the Big Woods Wildlife Management Area and State Forest, which someday should provide more habitat for the endangered woodpeckers. Biologists have successfully translocated red-cockaded woodpeckers to start new populations throughout the South. You could call it "assisted immigration," as young birds are captured from donor populations and then moved to new areas during the time of their lives when they would naturally disperse from the sites where they hatched.

As of 2018, there were no red-cockaded woodpeckers on Bill Owen's land, which now takes in around 1,850 acres. "We do have a few big trees here and there that are around 40 years old," he says. "I'm 64. There may never be any red-cockaded woodpeckers on this property in my lifetime, but I like to think that others will be able to see them here in the future."

In honor of his parents, Owen donated an easement on his land to The Nature Conservancy. The Nature Conservancy has named the preserve the Raccoon Creek Pinelands.

"Bill is a terrific example of a private landowner with a strong conservation vision for his property," said TNC's Brian van Eerden. "He's not only interested in protecting his land today but leaving a legacy for future generations."

soil's available plant nutrients, pH level, and characteristics.) If your property includes a forest elsewhere, you can identify the trees already growing there; depending on conditions on the area where you intend to plant, you may select the same species of trees, or choose other species better suited to the new site or to add diversity to your woods. To learn more about trees in your state or region, consult reference guides. For Pennsylvania, the Smiths suggest *Terrestrial and Palustrine Plant Communities of Pennsylvania*, by Jean Fike, produced by the Pennsylvania Natural Heritage Program, and downloadable for free on the internet. For Vermont, I would get down from my shelf a book titled *Wetland, Woodland, Wildland: A Guide to the Natural Communities of Vermont*, by Elizabeth Thompson and Eric Sorenson. Similar guides exist for other states. However, these are only general references. You must also consider your planting site's soil quality and moisture, elevation, and the degree of slope (if any) and its aspect (the direction the slope faces)—all of which can affect which kinds of trees will do best in that particular spot.

You can ask an independent professional forester, a cooperative extension wildlife or forestry specialist, or a service forester with your state's forestry or wildlife agency to visit your property and suggest possible trees for your site. These specialists can help you draw up a planting plan; if you go ahead with the project, they may follow up to make sure things go smoothly and help you solve any problems that arise. You can also contact your local office of the USDA Natural Resources Conservation Service, which may offer cost-sharing funds for purchasing and planting trees.

Along with choosing which species to plant, you need to decide on spacing. Planting trees on 20-foot centers requires around 100 trees per acre. You can also plant at 15x20-foot or 10x10-foot intervals. If you select a 20x20 grid, you will need 1,500 trees for a 15-acre project. You can often buy seedlings through sales put on by local soil conservation districts. Many states' forestry or wildlife agencies operate nurseries that sell seedlings to landowners for planting buffers around streams and ponds and creating or enhancing wildlife habitat. State nurseries use seeds from native stock and grow seedling trees in beds. Most of the seedlings they sell are one to several years old and six inches to a foot or so in height. They're usually removed from the soil while dormant and shipped in bags with no soil, and they are therefore known as "bare-root seedlings." Plant them using a spade or a planting bar. Another option is to use "containerized" stock, in which the seedlings' roots are enclosed by some amount of soil, depending on the age of the seedling and the size of the container. While these are more expensive than bare-root stock, they grow more quickly after they're put in the ground and are more apt to survive being transplanted.

Bare-root seedlings bought in quantity from a state nursery currently cost about 40 to 50 cents per tree (plus shipping), for a total of around $750 for 1,500 trees. Quantities of three- to four-foot trees will likely cost at least $5 to $6 per tree, for a total of $7,500 to $9,000. Will friends and family help you plant those 1,500 trees? If not, the NRCS or other professionals you've consulted can recommend a private contractor for the job.

It can be hard to accurately estimate the total cost of successfully establishing a woodland. A key factor is the number of deer in your area because these browsers can severely damage and suppress or kill recently planted trees. There are several strategies for planting small trees and getting them to survive in areas where deer are abundant.

You can build an eight-foot wire fence around the entire plot to exclude deer. Such fencing costs between $2 and $4 per linear foot for materials and installation. Fencing a square 15-acre plot will cost $7,000 to $14,000; it will cost more for a field with an irregular or elongated shape.

A second approach is to plant trees on 10-foot centers and let them fend for themselves, with the hope that 25 percent of them will survive and grow tall enough to get their branches, buds, and foliage above the reach of hungry deer. This approach quadruples the cost of the trees and the planting labor.

The third approach—and the one most people adopt—is to plant on a 20x20 or 20x15 grid and equip each tree with a plastic tube secured to a supporting stake. The tubes screen out sunlight except at the top, causing seedlings to put their energy into upward growth rather than sending out side branches. The tubes may deter deer to some extent; they also protect against mice, voles, and rabbits gnawing on the lower portions of the trees' trunks (if these animals girdle the tree by chewing off the bark around its trunk, the tree will die). A contractor will likely charge $12 to $15 per tree to supply the trees and to plant and protect them. For a "turnkey" job by a contractor putting in 1,500 seedlings, you'll pay around $18,000 to $22,500.

If you enroll in an NRCS conservation program, that agency may cover some of the project's cost. Also, your field may be eligible for their Conservation Reserve Enhancement Program (CREP), in which poor-quality farmland is taken out of production and turned into wildlife habitat; if so, NRCS will pay you a yearly per-acre rental fee that will help defray your overall costs.

Once the trees have been planted, your work is far from over. You should mow between the rows several times a year, both to maintain access to the growing trees and to control any invasive shrubs that might show up. For the first five or 10 years, you will probably want to mow around the trees in spring, mid-summer, and fall. Other chores include replacing trees that die, straightening and supporting ones that get bowed down by snow, or replacing a broken tree tube or resecuring one that has become detached from its stake. Broken limbs on the trees should be pruned, and

When trees are cut, more sunlight reaches the forest floor, giving lower plants and smaller trees a chance to thrive and often providing more food and cover for wildlife. If one of your forest management goals is to help wildlife, find a forester experienced in developing plans that integrate creating or improving habitat with producing timber.
Justin Fritscher, NRCS

branches growing the wrong way—back into the body of the tree—should be removed. Even after the trees have been growing for a few years, you need to monitor the ground beneath them for invasive plants such as thickets of honeysuckle or climbing bittersweet vines that may choke and overload the trees' branches.

Walking through a woods that you created yourself can yield a great feeling of accomplishment—especially when you see a bird's nest in the crook of a branch, watch a kinglet preening a tree's bark in search of dormant insects on a cold winter's day, or encounter a box or wood turtle lumbering along in the shade. But understand before you begin that 10- or 15-acre plantation that afforestation is a major effort, both in costs and the hours you will need to invest to make sure your project is a success.

CUTTING TREES

We started this chapter with the old adage that you can do three things with a piece of land: Plant something on it, cut something down, or do nothing. The second alternative, cutting something down, is probably easier to accomplish than planting and establishing a tract of woods from scratch.

Cutting trees, or logging, is often referred to by foresters and conservationists as "harvesting trees" or "harvesting timber." It may seem counterintuitive, but a well-thought-out, carefully conducted timber harvest usually will improve a forest's quality and health while also creating wildlife habitat. It's very important that you build a relationship with a professional forester who understands ecology and the habitat requirements of wildlife and who can help you decide on the best way to cut your timber, find a competent logger, and oversee a timber harvest so that it ends up helping rather than harming your woods. Be sure to ask a forester for references, and check them out along with his or her professional credentials.

Harvesting low-quality trees can shift growth into better-quality ones, gradually increasing the commercial value of the remaining timber in a *forest stand*, which is a localized group or community of trees that are uniform enough in species, age, and size that they can be distinguished from surrounding trees and managed as a unit.

A timber harvest can be used to diversify your woods. If you log, you may decide to cut common trees in an abundant age class, thereby promoting a broader number of species as well as trees of different ages. A diverse woodland will better withstand stresses such as diseases, insect outbreaks, droughts, and climate change as compared to woods in which the trees are almost all the same type and age. A diverse woodland will also provide habitat to a broader range of wildlife.

You can use timber cutting to increase the percentage or number of trees that offer high-quality food to wildlife: seeds, nuts, and fruits. Let's say your woodland is made up mainly of American beech, red and sugar maple, white ash, black cherry, and northern red oak trees. By conducting a timber harvest that focuses on cutting beech, maple, and ash, you can give the oaks and cherries a bigger share of the sunlight, letting them spread out their crowns into space formerly occupied by trees of the harvested species. The oaks and black cherries will gain vigor and health and produce more mast (acorns and cherries) for wildlife. Beechnuts are good wildlife food, too, so definitely retain some of them, knowing that beechnut crops are not as reliable from year to year as the oaks' acorns or the black cherries' fruits.

Harvesting Methods

There are many ways to harvest trees, but most foresters recognize two basic approaches: uneven-aged and even-aged harvesting. Both have their advantages and disadvantages, and both can be used to improve habitat for wildlife. If you want to make habitat for a certain guild, or group, of wildlife, make sure your forester understands and respects your intentions. Get him or her to fully explain all harvesting options before you sign a contract with a logger.

In an *uneven-aged timber harvest*, trees of different sizes and ages are removed through *selective cutting* on a management unit or across a forest stand. Foresters mark and loggers then cut single trees or small groups of trees. Uneven-aged management calls for more frequent harvests (known as *entries*) than does even-aged management. Conducted every five to ten years, selective cutting creates a number of small-scale disturbances throughout a forest stand. Sunlight can reach the forest floor in those places, letting nuts and other seeds germinate and tree seedlings and sprouts grow.

Uneven-aged harvesting maintains trees of many different sizes and ages in a stand. For that reason, some ecologists believe it comes closer to mimicking the natural growth and regeneration of forests than does even-aged harvesting. Selective cutting is often recommended for landowners who want to manage their woodlots or small forest stands for a steady flow of income without drastically changing the vegetation and tree species or the appearance of the woods. Uneven-aged harvesting may not be the best approach in areas where deer are abundant because the deer may browse off the limited number of seedlings and sprouts that grow where trees were cut, making it difficult or even impossible for the next generation of forest trees to arise.

Selectively cutting large oaks and hickories will reduce the number of nuts available to wildlife. Because it doesn't let in much new light, single-tree cutting favors seedlings of tree species that

Shade Tolerance of Eastern Trees

Shade-tolerant trees are able to grow and thrive in the shade cast by taller or competing trees or various terrain features. Shade-intolerant trees need full sunlight and few or no competing trees. Intermediate shade-tolerant trees fall between these two groups and can handle some shade. The information in this list comes from "Summary of Tree Characteristics," by Russell M. Burns and Barbara H. Honkala, in *Silvics of North America*, published by the U.S. Forest Service, and from other sources.

SHADE-TOLERANT

American Basswood, *Tilia americana*

American Beech, *Fagus grandifolia* (very tolerant)

American Holly, *Ilex opaca* (very tolerant)

American Hornbeam, *Carpinus caroliniana*

Balsam Fir, *Abies balsamea* (very tolerant)

Black Gum, *Nyssa sylvatica* (other *Nyssa* species are also shade-tolerant)

Black Spruce, *Picea mariana*

Boxelder, *Acer negundo*

Butternut, *Juglans cinerea*

Carolina Silverbell, *Halesia carolina*

Chinkapin Oak, *Quercus muehlenbergii*

Common Persimmon, *Diospyros virginiana* (very tolerant)

Eastern Hemlock, *Tsuga canadensis* (very tolerant)

Eastern Hophornbeam, *Ostrya virginiana*

Eastern Redbud, *Cercis canadensis*

Flowering Dogwood, *Cornus florida* (very tolerant)

Fraser Fir, *Abies fraseri* (very tolerant)

Green Ash, *Fraxinus pennsylvanica*

Laurel Oak, *Quercus laurifolia*

Loblolly-bay, *Gordonia lasianthus*

Northern Hackberry, *Celtis occidentalis*

Ohio Buckeye, *Aesculus glabra*

Red Mulberry, *Morus rubra*

Redbay, *Persea borbonia*

Shellbark Hickory, *Carya laciniosa* (very tolerant)

Slippery Elm, *Ulmus rubra*

Sourwood, *Oxydendrum arboretum*

Southern Magnolia, *Magnolia grandiflora*

Spruce Pine, *Pinus glabra*

Striped Maple, *Acer pensylvanicum* (very tolerant)

Sugar Maple, *Acer saccharum* (very tolerant)

White Basswood, *Tilia heterophylla*

Yellow Buckeye, *Aesculus octandra*

INTERMEDIATE SHADE-TOLERANT

American Chestnut, *Castanea dentata*

American Elm, *Ulmus americana*

Atlantic Whitecedar, *Chamaecyparis thyoides*

Bald Cypress, *Taxodium distichum*

Black Oak, *Quercus velutina*

Bur Oak, *Quercus macrocarpa*

Chestnut Oak, *Quercus prinus*

Cucumbertree, *Magnolia acuminata*

Eastern White Pine, *Pinus strobus*

Hackberry, *Celtis occidentalis*

Northern Red Oak, *Quercus rubra*

Overcup Oak, *Quercus lyrata*

Pignut Hickory, *Carya glabra*

Pumpkin Ash, *Fraxinus profunda*

Red Maple, *Acer rubrum*

Shagbark Hickory, *Carya ovata*

Southern Live Oak, *Quercus virginiana*

Southern Red Oak, *Quercus falcata*

Swamp White Oak, *Quercus bicolor*

Sweetbay Magnolia, *Magnolia virginiana*

Sycamore, *Platanus occidentalis*

Water Hickory, *Carya aquatica*

White Oak, *Quercus alba*

White Spruce, *Picea glauca*

Yellow Birch, *Betula alleghaniensis*

SHADE-INTOLERANT

Balsam Poplar, *Populus balsamifera* (very intolerant)

Basket Oak, *Quercus michauxii*

Bigtooth Aspen, *Populus grandidentata* (very intolerant)

Bitternut Hickory, *Carya cordiformis*

Black Ash, *Fraxinus nigra*

Black Cherry, *Prunus serotina*

Black Locust, *Robinia pseudoacacia* (very intolerant)

Black Walnut, *Juglans nigra*

Black Willow, *Salix nigra* (very intolerant)

Eastern Cottonwood, *Populus deltoides* (very intolerant)

Eastern Red Cedar, *Juniperus virginiana*

Honey Locust, *Gleditsia triacanthos*

Jack Pine, *Pinus banksiana*

Kentucky Coffee Tree, *Gymnocladus dioicus*

Loblolly Pine, *Pinus taeda*

Longleaf Pine, *Pinus palustris*

Mockernut Hickory, *Carya tomentosa*

Paper Birch, *Betula papyrifera*

Pecan, *Carya illinoensis*

Pin Cherry, *Prunus pensylvanica* (very intolerant)

Pin Oak, *Quercus palustris*

Pitch Pine, *Pinus rigida*

Post Oak, *Quercus stellata*

Quaking Aspen, *Populus tremuloides* (very intolerant)

Red Pine, *Pinus resinosa*

River Birch, *Betula nigra*

Sassafras, *Sassafras albidum*

Scarlet Oak, *Quercus coccinea* (very intolerant)

Shortleaf Pine, *Pinus echinata*

Slash Pine, *Pinus elliottii*

Sweet Birch, *Betula lenta*

Sweet Gum, *Liquidambar styraciflua*

Tamarack, *Larix laricina* (very intolerant)

Tulip Tree, *Liriodendron tulipifera*

Virginia Pine, *Pinus virginiana*

Water Oak, *Quercus nigra*

White Ash, *Fraxinus americana*

Willow Oak, *Quercus phellos*

tolerate the low-light conditions found beneath the forest canopy, including American beech, sugar maple, basswood, buckeye, eastern hemlock, balsam fir, and spruces. *Group-tree selection* will let in more light, opening up opportunities for trees that do well in partial shade, including black birch, hackberry, yellow birch, tulip tree, black cherry, and white ash. Oaks and hickories, which prefer more sunlight, will be less apt to thrive; the same goes for paper birch, aspen, butternut, and black walnut.

In an *even-aged timber harvest*, all the trees on a site are cut at the same time; they are then allowed to regenerate from their roots, stumps, or seed sources. This type of harvest leads to forest stands in which most or all of the trees are the same age, sometimes over a considerable area. The number of years between harvests or entries is known as a *timber rotation*. Depending on a site's growing conditions and the forest products that are harvested, time spans between entries can range from 20 years (for pulpwood and chips) to 75 to 100 years (for sawlogs).

Foresters generally use one of three methods to conduct even-aged timber harvests in eastern hardwood forests: clear-cutting, shelterwood cutting, and seed-tree harvesting. Clear-cutting is probably used most often; shelterwood and seed-tree harvesting are variations on the same theme. Ask your forester to describe the differences between these approaches, or learn about them in books such as Richard Yahner's *Eastern Deciduous Forest: Ecology and Wildlife Conservation* (University of Minnesota Press, 2000) and in publications produced by state forestry agencies and university extension services, many of which are available on the internet.

In general, for a timber harvest to qualify as a clear-cut, all mature trees within an area of at least one hectare (about two and a half acres) must be harvested. Foresters often recommend clear-cuts for sites or forest stands where most trees are of low quality, perhaps as a result of past "high-grading," when loggers cut the straightest, best, and most valuable trees, and left poorer-quality trees to continue growing. In such forests, it often makes sense to "wipe the slate clean" with a clear-cut and let a timber stand start over again.

Trees that need full sunlight will grow back quickly following a clear-cut. They include birches, white ash, black cherry, red maple, tulip tree, hickory, swamp tupelo, and southern pines. After clear-cutting, oaks and aspen grow back vigorously: aspen by sprouting from its extensive root system, and oaks by sprouting from their cut-off stumps and from acorns.

A clear-cut radically changes a forest's look and feel. The trees—all the trees—are gone. A recent clear-cut will provide habitat for some animals, and it will deprive other species of needed food and cover. For instance, if you love the flutelike calling of wood thrushes, you would not want to clear-cut much of your woods, since these birds need mature forest to breed. (Wood thrushes will, however, visit regrowing trees in clear-cuts to feed on abundant insects and fruits.) If you are fascinated by woodland salamanders, a clear-cut would not be for you, because the sunlight pouring down on the newly exposed forest floor will create warmer, drier conditions that make a habitat untenable for salamanders. If, on the other hand, you love seeing and hearing woodcock, conducting a series of clear-cuts on your lowland forest should attract these birds of young, regrowing forest. Ruffed grouse use young forests growing back on clear-cuts as well.

It makes sense to visit a clear-cut, check it out thoroughly, and then think hard about this harvest approach before you tell your forester to go ahead and mark out a timber sale. The Young Forest Project, a partnership among 17 states, along with federal agencies, timber companies, and wildlife and conservation agencies, has created thousands of acres of young forest to benefit wildlife, mainly through the use of clear-cut timber harvests. Check the Project's website to learn about young-forest demonstration areas where you can see both fresh clear-cuts and woodlands growing back after clear-cut logging.

Another important consideration before harvesting trees is the number of deer living on your land and on neighboring properties. In areas with high deer populations, you may need to put up expensive deer-proof fencing on a harvested tract to allow trees to grow and regenerate the forest stand. Or you may be able to "overwhelm" the deer by clear-cutting a large number of acres. Before you cut, get advice from a state-agency private-lands wildlife biologist or forester, a university cooperative extension forester, or a trustworthy professional forester with wildlife knowledge.

Planning a Timber Harvest

A landowner can use timber harvesting to create or enhance wildlife habitats on almost any property size. Let's consider a 20-acre tract. Over the course of 100 years, a landowner and his or her descendants could decide to harvest one-fourth of the tract (five acres) every 20 years. For each entry, cuts could be done on half-acre patches spread around on different parts of the property (perhaps keeping a grove of mature trees as a buffer around a house, cabin, or hunting camp). After five entries, one every 20 years, the entire tract would have been harvested one time, and during that 100-year period, wildlife would always have had access to many different age classes of trees.

On larger wooded properties—up to 100 or several hundred acres—a sequence of cuts can be set up to provide a steady stream of income to pay taxes or upkeep on a hunting camp or home while creating habitat for wildlife. Landowners may opt for uneven-aged harvesting. Or they may specify five- to 10-acre clear-cuts. Designing such clear-cuts with irregular shapes will help blend them into the landscape better than square or perfectly round cuts. (Competent professional foresters are very good at laying out clear-cuts in this manner.)

John Lanier is a habitat biologist who has worked for the U.S. Forest Service in the White Mountain National Forest in New Hampshire, for the New Hampshire Fish and Game Department, and for the Wildlife Management Institute. He notes, "A landowner who wants to help wildlife should try to create and maintain an even distribution of habitat components across the landscape—things like forest stands of different ages, old trees that have cavities used by wildlife, and areas where sunlight reaches the forest floor and lets low plants grow."

Whenever you log, preserve trees with cavities that wildlife, such as these young raccoons, use for shelter. Leaving patches of older forest in clear-cuts will encourage wildlife to venture farther out into openings created by timber harvests.
© *iStock.com/stanley45*

He offers this rule of thumb: "On every 10-acre clear-cut, leave about a quarter-acre of trees in a single, uncut patch. In that patch, make sure there's at least one potential cavity tree. What you're doing is creating islands of older forest in areas that will soon become young forest. By leaving those older forest remnants, you'll ensure that critters will venture farther out into the openings you've created." Don't be afraid to leave some single trees standing in the cut itself, ones that birds will use as perches from which to sing during the breeding season. Leave a few mature oaks and hickories to provide some acorns and nuts. You can also leave some healthy fruit-producing trees, such as black gums, hackberries, and black cherries.

Lanier points out that the types of trees on your land, along with local timber markets, can have a big effect on your harvesting options.

"Let's say you have 10 acres of mature woods," Lanier says, "and you want to cut an acre of trees to create young forest for wildlife, but not all in the same place—maybe you decide to make four quarter-acre patch cuts spread out over various parts of your property." Because many modern loggers use big (and expensive) machines, "they may not be attracted to such a small timber harvest, especially one for which they would need to move their machines four times. Instead of a high-volume commercial logger, you or your forester might find someone in the firewood business—one or two people with a pickup truck or a dump truck—to buy the trees and do the cutting. Or you might contact a small-scale or part-time logger who uses a tractor or horses to skid logs to a landing from which they can be loaded onto a truck and transported to a mill.

"Or let's say you have 50 acres and decide you want 10 percent of it to be young forest in the zero- to 10-year-old age class," Lanier says. "You would want to cut five acres every 10 years or so. Five acres might be large enough to attract a small logger, somebody with a cable skidder and a chainsaw rather than a big feller buncher and a grapple skidder. Horse- or tractor-logging would also be an option for a cut of this size.

"If you have 100 acres and want to keep a young-forest component on the land, then you will probably be cutting around 10 acres every 10 years. Your logging job is still on the edge of the small size—a logger with a $300,000 feller buncher might or might not be interested."

If your trees are all small- or medium-sized—not yet large enough for sawlogs—they may not have enough value for a logger to want to cut them. Same if they have a low market value because of their species—if they're not a type of tree, such as red oak, sugar maple, black cherry, or walnut, whose wood can be turned into high-value products such as furniture or flooring.

And if you live in an area where there are no professional loggers, and no mills to buy timber, you will probably need to search out a firewood cutter. This situation exists today in many areas where housing and other developments have changed the landscape from a rural one, with working forests and farms, to more of a suburban environment.

In 2017, this stand of northern hardwoods in Vermont had been growing for 34 years after being logged in 1983. Forests will grow back faster on better soils and in regions with longer growing seasons.

After You Log

At first, a timber harvest (especially an even-aged one, such as a clear-cut) looks messy—like a bad haircut or a recently picked cornfield. However, in just one or two growing seasons, the stumps, root systems, and nuts of harvested trees will start sending up thousands of new little trees to cloak the land. Light reaching the ground will spur the growth of food-producing plants such as blueberries, huckleberries, pokeweed, blackberries, wild strawberries, grasses, sedges, and wildflowers.

As it grows and greens up, the resulting young forest will provide food and cover to a wide range of songbirds, including many that nest in nearby older woods. Male American woodcock use newly logged sites as singing grounds during their springtime breeding displays. Areas of young forest often support healthy populations of small mammals, such as mice and voles, which attract predators. Cottontail rabbits, snowshoe hares, white-tailed deer, and black bears find food and hiding cover in the thick stems. Box turtles and wood turtles venture onto recently logged land to dine on fruits, seeds, and insects, and to lay their eggs in the sun-warmed soil. Many songbirds site their nests in areas of young trees up to 10 or so years old: golden-winged warblers, prairie warblers, yellow warblers, chestnut-sided warblers, field sparrows, indigo buntings, towhees, kingbirds, and dozens more. During migration, songbirds look for areas of young forest where they can rest safely in the thick cover and, in autumn, build up their fat reserves by eating high-energy fruits and insects

Deep-Woods Birds Need Young Forest, Too

Biologist Scott Stoleson works in the U.S. Forest Service's Northern Research Station in Irvine, Pennsylvania. Between 2005 and 2008, on four sites on the Allegheny National Forest, Stoleson used mist nets to capture 3,845 songbirds in mid- and late summer, after young birds had left the nest. Altogether, he and his colleagues caught and evaluated birds of 46 species. Of those, 45 percent were young-forest nesters, 22 percent were forest-edge nesters, and 33 percent were birds that nest in mature forest. Individuals of all 46 species were captured in regenerating clear-cuts—areas where mature trees had been harvested and where young forest was growing back. In contrast, only 29 species were netted in nearby mature woods.

Stoleson concluded that young forest in clear-cuts attracts and is heavily used by most forest-interior birds during the post-breeding season. The birds appear to boost their fitness by consuming the food resources amply present in the cuts, compared to birds that remained in deep-forest settings. Deep-forest nesters that Stoleson found using clear-cuts included scarlet tanagers, wood and hermit thrushes, veeries, red-eyed vireos, ovenbirds, redstarts, and blackburnian and cerulean warblers.

Birds that nest in mature forest, such as the rose-breasted grosbeak, often take their fledglings into young forest to feed on fruits and insects. The dense structure of the young trees and shrubs protects them from aerial predators such as hawks.
Ed Schneider

Stoleson's findings suggest that recent declines in the populations of forest-interior birds may stem in part from the increasing maturity and homogenization of our woodlands. He says, "Humans have really changed the nature of mature forests in the East. Natural processes that once created open spaces even within mature forests, such as fire, are largely controlled, diminishing the availability of quality habitat." He adds, "Some young-forest habitat within large areas of more-mature forest may be necessary to sustain some forest-interior bird species."

found among low plants and the dense stems and leaves of young hardwood trees.

Making a Scalloped Edge

A *scalloped edge* is more helpful to wildlife than a straight-lined edge because it adds more diversity to a habitat. Creating a scalloped or saw-toothed edge to your woodland is a tree-cutting practice usually done on a smaller scale than a timber harvest. You can make a scalloped edge wherever your woods meet a pasture, hayfield, or shrubland. Walk the edge of the woods and cut trees on the perimeter and a few yards back into the forest. Usually, the trees will be leaning toward the open space, since that's the side that will have more weight, because of the trees' branches growing out into the opening to intercept sunlight. Felling the trees into the field will make using their branches (for firewood) and their trunks (for firewood, pulp, or sawlogs) easier. Try to cut trees that don't provide top-quality food for wildlife while saving ones that produce more or better food, such as hackberry, black cherry, black gum, oaks, and hickories.

Confronting a Problem: Invasives

All too many forest stands in the East have alien invasive plants growing in their understory or midstory levels or along their edges. Autumn olive, ailanthus (also called tree of heaven), Oriental bittersweet, buckthorn, bush honeysuckle, barberry, kudzu, mimosa, multiflora rose, Japanese maple, Norway maple, and Japanese stiltgrass are just a few of the invasives that infest different parts of the eastern woods. Some of these plants reproduce vegetatively: Their root systems spread out and send up new shoots, which can spawn dense monocultures. Many release abundant seeds. Often they outcompete native plants because the invasives don't have natural pests or parasites to limit their growth and spread. Some of them release compounds from their roots that change the soil's chemistry, making it hard for native trees, shrubs, and ground plants to survive or establish themselves in areas taken over by invasives.

Recall that in Chapter 3 it was recommended that you thoroughly traverse your property to learn what plants and animals it supports. Early spring is a good time to visit your wooded acres and investigate whether invasives are growing beneath or among your trees. Many non-native species start greening up before our native species put forth leaves. Mark them for later treatment.

You can suppress or eliminate invasive trees and shrubs by using mechanical and chemical techniques. Cutting down invasives can slow their growth and limit their ability to produce seeds, but unless you cut them every year (and sometimes several times a year), this mechanical approach usually won't eliminate them. You may need to use chemical herbicides. A two-stage process works on most invasives: After they leaf out, cut or mow them, then, when they put forth new leaves, carefully apply a short-lived herbicide to the foliage. You can also cut them back and apply herbicide directly to the cut stems. It's a battle worth fighting to keep your woods healthy and better able to provide food and cover for wildlife. (See Chapter 9, "A Plague of Invasives," for more information.)

DOING "NOTHING"

Let's recall, one final time, that old forestry saying that you can plant something on a piece of land, cut something down, or do nothing. The last of those three options, doing nothing, can amount to simply waiting and letting the trees in your forest grow, while you enjoy hiking, snowshoeing, hunting, and watching wildlife among them. However, there are some very enjoyable and interesting things you *can* do to improve conditions for wildlife in your forest that have nothing to do with planting trees and only a little to do with cutting them down.

Nest Boxes

Nest boxes function like natural cavities in trees, features that are used by more than 35 kinds of birds and 20 mammals throughout the East. Nest boxes will be used by birds in forests, orchards, and small woodland clearings, on woodland edges, and in backyards. Resident birds such as chickadees, nuthatches, and brown creepers will tuck into them during winter's cold and snow. Mammals use them, too. I still remember how surprised and pleased I was to find four baby flying squirrels using a nest box I had built out of scrap wood and nailed to a tree on our old Pennsylvania homestead. That's one of the most intriguing things about nest boxes: You can't be completely sure who'll end up using them.

I still have the 1980s-era spiral-bound book *Woodworking for Wildlife: Homes for Birds and Mammals* that I bought for $3, complete with notes penciled in for the assorted boxes I built and put up on our land, both back in Pennsylvania and here in Vermont. Several states have produced versions of the book that can be downloaded for free on the internet. (Search on "wildlife nest box plans.") *The Audubon Birdhouse Book*, by Margaret Barker and Elissa Wolfson (Voyageur Press, 2013), is a good reference. You may also be able to get plans from your state's natural resources agency, your state university's cooperative extension service, and wildlife organizations such as the National Wildlife Federation and Audubon. If you don't want to build the boxes yourself, some state wildlife agencies and private companies sell pre-built ones.

You don't have to be a skilled carpenter to put together nest boxes, and wildlife won't care if they aren't perfect. Making and then putting them up in different settings can be a wonderful way to introduce a child or a grandchild to simple carpentry; they'll learn some important concepts about wildlife and will feel rewarded when they realize that they, themselves, can make a difference in the natural world—especially if they see wildlife using the boxes.

When building boxes, use three-quarter-inch pine boards (or whatever scraps of wood you have on hand), ring shank or coated nails (they hold the component pieces together better), and rust-proof hinges. Paint the boxes some dark color to protect the wood, or let them weather to a dull gray. Drill holes in the bottom for drainage.

Boxes of different sizes and styles can accommodate wood ducks; screech, saw-whet, barred, and great horned owls; kestrels; a range of woodpeckers from the diminutive downy woodpecker to the big pileated woodpecker; chickadee, tufted titmouse and white- and red-breasted nuthatch; wrens; bluebird; tree swallow; and prothonotary warbler. Starlings and English sparrows are cavity nesters that most people have no desire to attract, but they use nest boxes, too, sometimes usurping them from native birds. Open-sided structures will attract phoebes and robins. Barn owl boxes can be erected on high beams inside barns. Plans are also available for boxes that bats, squirrels, and raccoons will use for nesting, resting, and surviving extended spells of harsh weather.

Preserving and Creating Cavities

Woodlands 50 years and older often have trees large enough to have developed natural cavities. Some of these may be "wolf trees"—trees that, many years earlier, were passed over by loggers because they were already hollow or were too twisted or crooked to be worth felling and hauling to the mill. Some of these might be old shade trees left for livestock on hillside pastures, or, in flatter terrain, for plowmen and their teams; as decades passed, trees took over the old pastures and fields, and regrowing forests swallowed up these antiquarian survivors.

Snags—dead standing trees—are another source of natural cavities, as are stubs, which are dead trees whose tops have snapped off anywhere from a few feet to many feet above the ground. You can create snags by "girdling" living trees: using an axe or a chainsaw to cut a groove or notch in the bark, half an inch deep or deeper, all the way around the tree's trunk. For a trunk diameter of 18 or more inches, the groove should be six to eight inches from its top edge to its bottom edge. This breach stops the flow of sap, causing the tree to die within a few years. Later, fungi will enter the dead tree in various places, and the ensuing internal decomposition will either create cavities or soften the wood so that animals can more easily make holes and excavate chambers. Stumps and logs lying on the ground can have cavities, too.

Snakes and tree frogs take shelter in cavities. Chimney swifts and bats roost in hollow trees in forests and fencerows. Squirrels, mice, raccoons, opossums, skunks, porcupines, mink, martens, fishers, and gray foxes are some of the other creatures that use cavities. Black bears hibernate inside large hollow trees. According to the Kentucky Department of Fish and Wildlife, "[Wildlife] species that use smaller snags may actually prefer the larger ones. There is no specific number of snags that should be maintained or produced because the optimum number of snags required by different wildlife species varies. Generally, six snags per acre could be considered an absolute minimum while as many as 30 per acre is an optimum objective."

Standing dead trees, known as snags and stubs, are a great source of holes and hollows used by reptiles, amphibians, birds, and mammals. Never cut cavity trees for firewood, and preserve them during timber harvests.

Because they're so valuable to wildlife, cavity trees should never be cut for firewood. It is much better to obtain firewood by thinning trees that stand too close together or to favor a type of tree that produces high-quality wildlife food over another that doesn't. (If you're working to increase the value of the timber in your woods, cut crooked trees or ones that fork close to the ground.) The only reason to cut down a cavity tree is if it overhangs and may fall on a frequently used road or hiking path or a building.

Tending Toward Old Growth

Very little true old-growth forest survived the rapacious logging that swept across eastern North America from the 1800s into the early twentieth century. Stands of old-growth trees remain in many states, and they are wondrous places to visit—they are also poignant when you see what was lost when essentially all of our primal natural forests fell to the axe and saw.

Healthy old-growth forests have multiple layers of vegetation from ground level up to the canopies of the tallest trees, which may be over 100 feet above the ground. They also have a beautifully variable pattern, with different densities and ages of trees across a stand, gaps in the canopy where trees have died or been knocked over by high winds, younger trees rising up to fill those gaps, abundant standing dead trees, rotting trunks on the forest floor, and cavities in both living and dead trees used by wildlife. In old-growth forests there exist tremendously complex (and only partially understood) webs linking trees of the same species, as well as trees of different and competing species. Vast underground networks of fungal mycelium meld with the roots of trees and shrubs, helping them take in nutrients and perhaps even communicate with one another. *The Hidden Life of Trees*, by Peter Wohlleben (Greystone Books, 2016), is a highly readable book that opens a window on this fascinating aspect of the natural world.

Mosses and lichens grow widely throughout old-growth stands. Colonies of spring wildflowers, such as trilliums, spring beauties, Dutchman's breeches, trout lily, and hepatica, carpet the forest floor. Decomposing tree trunks and branches lying on the ground—called "coarse woody debris" by ecologists—provide habitat for invertebrates such as mites, beetles, centipedes, slugs, and spiders. Salamanders abound, as do snakes, shrews, mice, and flying squirrels—these, in turn, feed numerous predators higher up on the food chain.

Today, as forests in the East grow older, they are beginning to show some of the complexity and characteristics of old-growth stands. It will take many years before woodlands can rebuild many of the complex life webs that characterized the original forests. Can true old-growth forest ever return? It's a tough question to answer. Some key species of trees are no longer with us: American chestnut, felled by an imported blight, is essentially gone from the eastern woods; hemlocks are under siege from the hemlock woolly adelgid; more and more beech trees fall ill with a bark disease caused by a scale insect's feeding combined with a deadly fungus; and the emerald ash borer has killed ash trees by the millions.

If you as a landowner are lucky enough to be the temporary steward of a late-successional forest—one that is, say, 80 to 100 or more years old—there are a number of things you can do to speed up the rate at which your woodland habitat matures.

Bill Keeton is a forest ecologist with the University of Vermont. He has come up with a method he calls "structural complexity enhancement," which he is testing on study plots in different parts of Vermont. In summer, he supervises field crews who go out and collect data from trees, plants in the understory, and soils; these results are compared to plots where more conventional forest management approaches are being used.

Professor Keeton has said that his technique "emulates the natural disturbance processes that gradually encourage forests to develop late-successional characteristics." It includes harvesting trees but taking about 40 percent less timber volume than through conventional cutting regimes; leaving many of the largest trees uncut; creating gaps in the forest canopy around some trees, so that their crowns expand rapidly to fill the openings and allowing them to get bigger faster; cutting down some trees and leaving them on the ground to rot; girdling trees to make snags; and pushing trees over so that their roots get yanked out of the ground, creating the "pit-and-mound" terrain that characterizes many old-growth stands. Keeton makes sure that trees of many different sizes are left following any harvesting, maintaining complexity in the forest while increasing biomass. He believes the technique fosters resilience in forest stands against drought, higher temperatures, and disease.

It also increases the forest's ability to store carbon in its trees and soils. Keeton avers that large landowners who use the technique—ones who own 1,500 acres or more—can sell carbon credits while still making money from harvesting and selling timber. At the same time, they can build a healthier forest and restore old-growth habitats. On his study plots, 10 years after harvesting, carbon storage was just 16 percent lower than it would have been in forest stands where no cutting took place. (Some conventionally harvested stands retained 45 percent less carbon than did the uncut stands.)

Keeton's approach may interest timber investment management organizations, or TIMOs, which help investors, including institutions such as universities, actively manage forested lands to provide income. It may also appeal to land trusts, nature organizations, municipalities with stewardship responsibilities for town or county forests, and other groups that manage large acreages of forest. For a person who owns a sizable tract of trees that are 100 or more years old, adopting some of Keeton's methods could be a fascinating project.

5

Grasslands for Wildlife

THERE'S SOMETHING ABOUT A GRASSLAND that attracts us in a visceral way. Maybe it's the shimmer of light on blades and stems or the way the wind combs through the grasses and other plants. Perhaps it's an instinctive sense that we are safe in an open setting where we can see all around—even though threats from predators no longer exist in our modern world. Or maybe it's simply an understanding that natural grasslands are rich, vibrant places where a close encounter with a wild creature is always possible and often imminent.

Many kinds of wildlife find food and cover in grasslands. Meadow voles and meadow jumping mice eat the stems, blades, and seeds of grasses and associated broadleaved plants. In summer, cottontail rabbits and white-tailed deer emerge at dusk to feed on the vegetation. Kestrels, hawks, shrikes, owls, weasels, red foxes, and coyotes prey on insects, amphibians, birds, and small mammals

Conservationists planted native warm-season grasses in this 25-acre field on a Connecticut wildlife management area. Landowners can make similar plantings for wildlife.
Judy Wilson, CT DEEP

LANDOWNER STORY

Bringing Bobolinks Back to New England

Bobolink males have striking black, white, and yellow plumage and sing an intricate, bubbling song. Females, more drably colored, site their nests on the ground in grasslands. © *iStock.com/mirceax*

KARAN AND STEVE CUTLER'S BACKYARD BORDERS the shore of Lake Champlain in western Vermont. "Living so close to the lake," Karan said, "we see a good variety of bird life throughout the year—waterfowl, raptors, shorebirds, and songbirds. We've been lucky to have nesting ospreys every year since Green Mountain Power put up a platform in our field."

What the Cutlers didn't see or hear many of until a few years ago, even though their property includes a 10-acre hayfield, were grassland birds, including bobolinks, whose males have striking black, white, and yellow plumage and sing an intricate, bubbling song.

Bobolinks winter in South America. In spring, they migrate to breeding grounds from roughly Pennsylvania to Nebraska and north into southern Canada. In times past, they nested in tallgrass and mixed prairies; today they also use hayfields and meadows. Bobolinks nest on the ground. They need at least 10 acres of grassland to breed successfully—and a greater area is even better. The colorful males stake out territories and compete for the attention of the drab females, which look like oversized sparrows. Nesting and egg-laying commence in mid- to late May in

Vermont. The young hatch in mid-June—about the same time that farmers start cutting hay.

Nests get wrecked. Broods are destroyed. Repeated cuttings keep bobolinks from successfully renesting and rearing young. This problem exists across the species' breeding range. According to the North American Breeding Bird Survey, a monitoring program run jointly by the U.S. and Canadian governments, the continent's bobolink population has fallen by more than 2 percent annually since 1966, for a cumulative decline of 65 percent.

It turns out that there's a way for farmers to continue to make hay and for bobolinks to breed. That process drives the Bobolink Project, sponsored by the state Audubon chapters of Vermont, New Hampshire, Massachusetts, and Connecticut. Advice from Margaret Fowle, a biologist with Audubon Vermont, led Karan Cutler and three of her neighbors to sign up for the Bobolink Project.

"Probably none of us would have qualified by ourselves," Cutler said, "because our individual fields aren't large enough." But together, those four properties amount to 35 acres of connected hayfields—large enough for bobolinks to use, along with other grassland birds such as eastern meadowlarks and Savannah sparrows.

"When we moved here around 10 years ago, we saw very few bobolinks," Cutler said. "Now we see them throughout the summer."

The Bobolink Project provides mowing guidelines based on studies of bobolink nesting patterns by researchers with the state universities of Rhode Island, Connecticut, and Vermont. Essentially, the Project pays farmers either to delay mowing until after the nesting season or to mow early and then hold off on mowing again for at least 65 days, giving grassland birds a window of time during which they can nest and bring off broods. Private donors contribute funding that's held in a pool. (Learn more about the Bobolink Project by visiting the organization's website, where you can make a financial donation to advance this inventive conservation effort.) Landowners apply for the funds, and natural resource professionals such as Vermont Audubon's Fowle evaluate their properties, determine whether bobolinks are likely to use the grasslands, and write a management plan that includes a bobolink-friendly mowing schedule.

"Applying to the Bobolink Project was easy," Cutler reported. "I was able to do it all online."

Tim Howlett of Champlainside Farm mows the Cutlers' and their neighbors' fields and uses the grass to feed his dairy herd. Now he's on a new schedule: He can mow before June 1, but then he must wait until August 15 to cut again. While some bobolink nests are destroyed by that early first mowing, the birds nest again, and the young in the second brood have enough time to develop and fledge before the mowing machines return. The Bobolink Project sends Cutler and her neighbors a check, and they turn the money over to Howlett. He gets only two cuttings of hay off those 35 acres, instead of his usual three, but the payment makes up for at least some of the difference.

"This complex of fields makes a nice habitat," said Fowle, of Audubon Vermont, as she and Cutler strolled along a gravel road—Cutler's driveway—that curved through the hayfield's knee-high grass. It was a bright, sunny day in late July. Monarch and swallowtail butterflies fluttered above the fields. "The grass doesn't grow that tall here, but the bobolinks do very well," Fowle said. "They'll tolerate a certain number of forbs [flowering plants], although they prefer grasses. They don't nest in fields with shrubs."

A female bobolink builds a nest by clearing a tiny patch of bare ground, often at the base of a larger meadow plant such as a red clover, then weaving a small cup out of grass blades and stems. She lays three to seven eggs in the nest and incubates them for 10 days to two weeks. After the young hatch, both parents feed them high-protein insects. Ten or so days later, the young birds fledge, leaving the nest and—as they are not yet able to fly—walking around in the grass and learning to catch insects on their own. They do best in fields where the grass isn't too thick since they need to be able to move about between grass clumps when foraging. After a few more days, their flight feathers grow out, letting them fly to other cover when the next round of mowing begins.

Fowle trained her binoculars on a bird flitting about 10 feet above the ground. It flew in a big circle, and then landed on a tall weed stalk. It was a bobolink, although not a male in breeding plumage: This bird was a tan color with stripes on its wings, so it was either a female or a male in post-breeding plumage. It darted its head and caught a small green caterpillar in its beak.

"This spring, when I walked transects through the fields, I counted 20 female bobolinks," Fowle said. "I also saw meadowlarks here this summer—they need a minimum of around 20 acres to breed."

Vermont Audubon biologist Margaret Fowle observes young bobolinks fledged in a field where delayed hay mowing let adult bobolinks successfully raise broods. The Bobolink Project is sponsored by state Audubon chapters in New England.

Cutler raised her head as a half dozen other bobolinks, all in drab plumage, flew into the field and landed on weedstalks. "Two weeks ago," she said, "the neighbor saw about 45 bobolinks. Savannah sparrows are here, too."

Fowle noted that the birds would soon gather into flocks and get ready to migrate south.

Bobolinks move by stages down to Florida, then cross the Caribbean, stopping on islands before reaching northern South America and ultimately wintering even farther south, on the Argentine Pampas. Coming back north in spring, males and females often return to the places where they bred successfully the previous year. Yearling birds generally breed within a mile or two of the areas where they hatched. Establishing a population here on the eastern shore of Lake Champlain will mean more birds returning to the same area in years to come. In 2017, Audubon Vermont worked with 15 landowners to set up delayed mowing on 600 acres of grasslands—up from previous years' totals of around 500 acres. An estimated 229 pairs of bobolinks nested on those managed grasslands, producing a record number of young: Fowle and her fellow Audubon biologists estimated that some 639 chicks survived to fledging.

"We want to do everything we can to keep this open, grassland habitat on our land," Cutler said. "You see lots of hayfields in the Champlain Valley, but most of them are mowed early and often, so they're not as congenial to wildlife." She reflected on being part of the Bobolink Project: "It's been fun. Friends who are birders exult when they come here and see bobolinks. I'd like to think we've made a difference."

drawn to these open, light-drenched settings. Wild turkeys feed on insects and seeds, as do ring-necked pheasants and bobwhite quail. In small grassy openings—such as log landings that have been seeded with grasses following timber harvests—ruffed grouse and their broods find high-protein insects for food. Male American woodcock claim grassy openings in early spring; at dawn and dusk, they sound a buzzing *peent* call from the ground, and then they fly up into the air and come spiraling down again, singing a beautiful flowing song to attract females.

Insects abound in grasslands: grasshoppers, crickets, dragonflies, bees, and butterflies such as swallowtails, monarchs, and fritillaries. Caterpillars of the endangered Karner blue butterfly feed on wild blue lupine, a flowering plant found in some grassland habitats. Bats hunt for flying insects above grassy openings, as do swallows, kingbirds, phoebes, olive-sided flycatchers, whippoorwills, and nighthawks. Amphibians and reptiles that use grasslands include frogs, toads, turtles (box, wood, spotted, Blanding's, and others), and snakes (hognose, garter, green, black racer, massasauga, and more). Male songbirds perch on upright stems, singing to stake out nesting territories or attract a mate or scanning their surroundings for insects or other invertebrate prey to feed themselves or their young. Some birds nest on the ground in grasslands, including upland sandpiper, grasshopper sparrow, vesper sparrow, Savannah sparrow, field sparrow, song sparrow, dickcissel, meadowlark, bobolink, northern harrier, and short-eared owl.

If you have a yard, you can convert part of it into a small natural grassland that will require much less maintenance—and will provide more food for wildlife, and more enjoyment for you and your family—than the groomed, cosseted grass of a conventional lawn. If your property includes farm fields, you can limit the amount of mowing that you do and let existing grasses and wildflowers grow tall and thick. To create a high-functioning grassland habitat, you can plant a meadow, field, or prairie of native warm-season grasses.

GRASSLANDS THROUGH THE CENTURIES

It's a common belief that before Europeans arrived, eastern North America was one vast forest. That's not the case. Although most of the region was wooded (we don't know the exact percentage), openings existed among the trees. Such openings included areas where beavers' dam-building activities killed trees and where grasses, sedges, and wildflowers seeded in; fields cleared by Native Americans for growing crops, and areas they burned to stimulate the lush growth of low vegetation, drawing in animals they hunted; grasslands that sprang up in the wake of lightning-caused fires; and grasses and heath plants (shrubs such as blueberries and cranberries) covering places where the soil was too thin, sandy, or acidic for trees to grow. Farther west, the forests thinned out, and tallgrass prairies dominated. In the South were many grasslands, including ones beneath and among the longleaf pines common in that region.

Openings in the primal forest attracted European settlers. My eight-greats grandfather, Nathaniel Foote, is said to have been one of ten "adventurers" who, in 1634, walked from Watertown in the Massachusetts Bay Colony west to the Connecti-

Wildlife use grass fields for many different life functions. Many animals eat grasses and low broadleaf plants or use the vegetation as cover. This whitetail doe feeds on grass as her young fawn waits to nurse.
Laura Jackson

cut River, where they bought previously cleared land from the local Indians. Nathaniel and his fellow settlers, Puritan immigrants from England, would have called their purchase "champion"—their word for flat, open land without trees or hills, eminently suited to farming. A book, *The History of Ancient Wethersfield, Connecticut*, says that the local Wogunk Indians had their own word for such land: *pa-qui-auke*.

In the centuries following European settlement, the eastern forest was largely cleared, the wood used for fuel and a range of products, and much of the land was converted to agriculture. Later, many acres of farmland were abandoned as people moved to regions with more-fertile soils or gave up on farming and relocated to cities and towns. Gradually, forest reclaimed those open acres. (In many parts of the East today, moss-covered stone walls cutting through the woods show where people once grew crops or pastured sheep and cattle.) Since open areas are easy to build on, housing and commercial development claimed thousands of acres of former farmland, including many grasslands.

Today, grasslands of differing qualities and usefulness to wildlife can be found on working farms and recently abandoned farms; on utility and railroad rights-of-way; on reclaimed surface mines; along the sides of rural roads; in old cemeteries; and in fields planted by state and federal wildlife agencies. Recreation fields, corporate parks, airports, and landfills also support grasslands. Some of the most extensive and productive grasslands for wildlife exist on military bases.

Two excellent books will educate you about grasslands, their history, and their value to wildlife and the environment: *Forgotten Grasslands of the South: Natural History and Conservation*, by Reed F. Noss (Island Press, 2013), and *Grasslands of Northeastern North America: Ecology and Conservation of Native and Agricultural Landscapes*, by Peter D. Vickery and Peter W. Dunwiddie (Massachusetts Audubon Society, 1997).

TWO TYPES OF GRASSES

Two general types of grasses grow in the East: cool-season grasses and warm-season grasses.

Cool-season grasses are not native to the region. Widely planted for grazing purposes, these European grasses include smooth bromegrass, tall fescue, smooth meadow grass (also called Kentucky bluegrass), timothy, and orchardgrass. Alfalfa and clover are non-native legumes that are often planted along with the cool-season grasses. Cool-season grasses green up and start growing early in the spring when air temperatures reach 41 degrees Fahrenheit. Livestock readily graze cool-season grasses, which can also be cut for hay and silage. Cool-season grass mixes are often planted for lawns.

These grasses require ongoing maintenance that can include adding fertilizer and lime, using herbicides to kill competing plants, and occasionally reseeding stands. Farmers may take hay off cool-season grass fields three or four times in a single growing season, which doesn't give grassland-nesting birds such as meadowlarks and bobolinks enough time to build their nests and then incubate, hatch, rear, and fledge their young. When cool-season grasses are allowed to grow tall, as in old fields that are infrequently mowed, they may grow so densely that songbirds, rabbits, quail, and other wildlife cannot move easily between the stems.

Stands of cool-season grasses may have few other plants mixed in. (Tall fescue even produces a toxin that inhibits the growth of other plants, including native ones that are becoming increasingly rare.) Less plant diversity means fewer butterflies, bees, moths, birds, and small mammals. Cool-season grasses can get knocked down by snow, so they may not offer good cover to wildlife in winter and early spring.

Warm-season grasses include switchgrass, Indian grass, big bluestem, little bluestem, and broom sedge. These native grasses once covered the prairies of the Great Plains and grew in savannas

Sometimes called "bunch grasses," native prairie plants once grew in savannas and glades in the East. Landowners can plant patches of these low-maintenance warm-season grasses, which can grow as high as four to six feet, with fairly open ground between clumps of individual grass plants.

and glades that accented the eastern forests. They are known as "bunch grasses" because the individual grass plants grow in bunches or clumps. Warm-season grasses green up later in the spring than do cool-season grasses, starting their growth when air temperatures reach 60 degrees Fahrenheit and soil temperatures reach 50 degrees Fahrenheit. They continue growing during the heat of midsummer when the growth of cool-season grasses slows. They rarely need fertilizer or lime. About the only maintenance they require is mowing or burning every few years to knock back any trees or shrubs that may invade a field and to remove some of the accumulated residual growth and plant matter.

The extensive root systems of warm-season grasses spread wide and grow deep, penetrating the soil five to six or more feet. Their roots help warm-season grasses take in water during drought, efficiently pick up nutrients, and grow back quickly following a fire. The roots completely regenerate every three to four years, adding organic matter to the soil and boosting soil fertility. They also anchor the soil, reducing erosion and runoff into rivers and streams.

Warm-season grasses typically grow in upright clumps with bare or lightly vegetated ground beneath and between the individual grass plants. Ground-dwelling birds such as quail take dust baths in exposed soil in warm-season grass patches.

They can move freely between the clumps with their young; the grass stems arching overhead help shield them from predators. Rabbits move about using these spaces, too, and white-tailed deer sometimes hide their young fawns in warm-season grasses. The bunching growth habitat lets wildflowers and other herbs (non-woody plants whose above-ground parts die back at the end of each year) grow between the grass clumps. The resulting diversity of vegetation attracts a broader range of insects, creating more and better feeding opportunities for wildlife. Since the sturdy, rigid stems of warm-season grasses resist bending over and matting down under snow, they offer shelter to wildlife in winter and provide ready nesting cover in spring. If heavy snow does weigh the grasses down, they will often spring back up once the snow gets blown off or melts.

Warm-season grasses look very different from the grasses in hayfields and lawns. They're much taller and have handsome fountain shapes. Big and little bluestem were both named for the bluish-green hues of their blades; in autumn, they turn reddish-purple. The seed heads of little bluestem look like tiny tufts of cotton. Other colors adorning native grasslands include brilliant greens in spring and warm golds and russets in the fall. If you mix in wildflowers when creating a warm-season grass meadow, the color palette will be even more varied. In winter, the textures of warm-season grasses are attractive and eye-catching.

MANAGING, ENHANCING, AND CREATING GRASSLANDS

In many regions, the only remaining grasslands are the hayfields on working farms. Conservation-minded farmers can set up a rotational mowing or grazing program so that different parts of a field are mowed or grazed at different times. This strategy leads to a patchwork of varying grass heights, providing food and cover to the broadest range of wildlife. Keeping some areas of ground exposed benefits killdeer and horned larks, which site their nests on bare soil. Farmers can raise their mowing bars to six inches or higher in areas where grassland birds nest. Since these birds roost in fields at night, don't mow after dark.

For fields where intensive grass or hay production is not the top priority, farmers can hold off on mowing until after August 1, which is when grassland birds typically have finished breeding. Other areas where delayed mowing can help wildlife are airfields, landfills, fallow fields, marginal farmland, weedy areas, and fields used to produce bedding straw. Federal and state cost-share programs may help pay for delayed mowing and other farming practices that benefit wildlife. County extension agents and the USDA Natural Resources Conservation Service will have information. One grassland-oriented conservation program is the Bobolink Project, in which farmers receive funds to delay mowing. Even if you don't own a grassland, you can help by making a donation to the Bobolink Project. (See the "Landowner Story" in this chapter.)

I was only vaguely aware of native warm-season grasses when my family and I moved to northern Vermont in 2003. If I'd known more about them, I might have considered converting two old fields on our land, each about two acres, to warm-season grasses. (Or, since we had just spent gobs of money fixing up an old house, economics might well have kept me on the course I followed: cutting down small trees that had invaded the fields and letting the existing cool-season grasses and flowering plants continue to grow between scattered thornapple shrubs and apple trees.)

I already described how I improved those two old fields for wildlife in the first chapter of this book so I won't go into as much detail here. By removing the aspen, red maple, black cherry, and white ash trees that had seeded into the fields, I stopped the forest regeneration process that gradually would have turned those two grasslands back into woods.

A tiger swallowtail takes nectar from a hawkweed. Ruffed grouse, wild turkeys, and goldfinches eat seeds of hawkweeds, and rabbits and deer nibble on the foliage. Most of the commonly seen hawkweeds are not native to North America.
Tom Berriman

The fields include timothy and orchardgrass, along with red clover. (I don't know when the fields were planted, but this place has been a farm since the mid-1800s.) Over the years, native and alien wildflowers have gained a foothold: violets, hawkweed, buttercups, cinquefoil, wild strawberry, aster, chickweed, dandelion, daisies, black-eyed Susans, Queen Anne's lace, goldenrod, and milkweed. I have never tried to identify and list all the plants mingling in our meadows; dozens more species likely are present. I have found two aggressive invaders. One is bush honeysuckle, a quick-growing shrub. In the spring, when the soil is damp and when the honeysuckles put out leaves in advance of native plants, I locate and pull out the smaller honeysuckles, getting as much of their root systems as possible. I use loppers or a walk-behind brush hog to cut back the larger honeysuckles. A second invasive in the two fields is smooth bedstraw, a low-growing perennial weed that produces many small seeds. Short of using herbicide on the whole field and replanting it, I don't know how to eliminate the bedstraw. I can't really attack it through mowing because it mingles with many "good" plants. Fortunately, it remains a fairly minor component in the two fields.

What I've done is basically kept the fields functioning as grasslands habitat—not a difficult task (except for getting stung now and then by yellow jackets that take umbrage at my walk-behind mowing machine trundling over their in-ground nests). In the past, I have cut the fields every second or third year, waiting until after August 15 when birds such as song and field sparrows have finished nesting. Nowadays, I mow even later; milkweeds, goldenrods, and asters flower in late summer, and I don't want to cut down those plants and deprive monarch butterflies of an important nectar source before and while they're migrating south. Next year I plan to mow in early spring, before any birds start nesting, cutting back the small trees that continue to sprout in those two pocket grasslands.

A friend of mine has chosen a similar strategy for the field around his house in central Pennsylvania. Randy Hudson is an architect and a landscape designer. Around 25 years ago he and his wife, Cynthia Nixon, built a beautiful house (Randy designed it) on an 8.5-acre parcel, part of a farm that had been divided up into large lots. They started off with corn stubble in an otherwise bare field around their house. (That would have been an ideal setting in which to plant warm-season grasses—but they, like me, were unaware of that option.) Thinking they might keep horses, Randy and Cynthia hired a farmer to plant a commercial pasture mix of cool-season grasses. Horses never became part of the program, and since the cool-season grasses were only mowed once a year for hay and were not fertilized to maintain their vigor, weeds and wildflowers began appearing. Over time, Randy and Cynthia noticed how many wild animals were using the field—bluebirds, red-winged blackbirds, butterflies, and bees—so they decided to stop haying the field altogether and see what would happen.

Near their house they had planted evergreen trees to screen views of other houses and to make a year-round shelter for birds. They also put in over a hundred food-producing native trees and shrubs. In the field, goldenrod and milkweed

Milkweeds, goldenrod, and asters are important late-summer nectar sources for butterflies, including monarchs as they migrate south. These perennial plants thrive in old farm fields.

compete with invasive Canada thistle, crown vetch, bindweed, and quack grass. "I try to mow back the invasives since they don't offer good food for the birds and insects," Randy says. He also mows paths through the field and around its perimeter, so that he can easily access different parts of the field to see what's blooming and which animals are about—and sometimes just to carry a lawn chair out into the meadow and sit watching and listening or seeing the sun go down.

"It's amazing how great this works," Randy says. "Now we see clouds of monarch butterflies—and bees, too—every summer." To keep trees out, the fields are mowed in late summer, after the birds are done nesting. It's clear that Randy and Cynthia get a lot of satisfaction out of a field full of unruly plants, along with its many wild inhabitants. In 2017, the National Wildlife Federation cited the couple for their participation in NWF's Garden for Wildlife Program, recognizing their grassland as a Certified Wildlife Habitat and part of their Million Pollinator Garden Challenge, a national effort to restore habitat for pollinating insects.

PLANTING GRASSLANDS

The kinds of wildlife that will use a grassland vary with the size of the tract and the types of habitat surrounding it.

If you own a wooded area and create within it a one- or two-acre patch of grasses, you will almost certainly bring in deer and wild turkeys. Depending on where you live, you may attract quail or ruffed grouse and their broods. Medium-sized predators and omnivores such as coyotes, foxes, skunks, and raccoons will likely use the habitat. In spring, black bears may graze on the grasses' tender new growth. Box turtles, green snakes, and garter snakes will hunt for insects and other invertebrates; larger snakes, such as black rat snakes, will prey on small mammals. A wide range of forest birds will seek out grassy openings in otherwise-wooded settings since they produce more and different insects than do the neighboring woods.

If, on the other hand, your grassland is a large one bordered by open country that includes other grassy habitats, you may attract specialized grassland birds (biologists call them "grassland-obligate species") that need sizable expanses for breeding. These birds include bobolinks, which nest in fields 10 acres or larger, and meadowlarks, which require fields 20 acres or larger. Large grassy expanses can provide nesting habitat for short-eared owls and northern harriers, two birds whose populations have fallen steeply in recent decades as grasslands have become fewer, smaller, and more fragmented. These two raptors subsist on small mammals, such as meadow voles, that live in grasslands. The barn owl is another bird that finds its prey in grasslands. In winter, northern birds—seed-eaters such as horned larks and snow buntings, as well as predators such as rough-legged hawks and snowy owls—gravitate to grasslands, which share many of the characteristics of the arctic tundra where these birds breed in summer.

Whatever the size of your project, rather than planting a pure stand of cool-season or warm-season grasses, add some wildflowers if your goal is to help wildlife. Nurseries and seed suppliers offer seeds for a wide variety of native wildflowers that are easy to grow, drought-resistant, and colorful. Perennials do not need to be replanted, as they come up from their root systems each year; annuals often need replanting. Remember, diversity is the name of the game. The more different nectar- and seed-producing plants you add to a grassland, the broader the range of insects and wildlife it will support. Consider planting some cool-season grasses along with warm-season types. Again, the diversity will attract and sustain wildlife. If you have enough land, you can plant blocks of warm- and cool-season grasses next to each other.

A small patch of warm-season grasses mixed with wildflowers can beautify the area right around your house. Be sure to use native species. Check the Audubon Society's Native Plant Database online and learn which are the best plants for birds in your area. Local Audubon chapters often put on native plant sales in spring; staff members may be able to give you personalized advice. (Learn more about making a small planting of warm-season grasses and wildflowers in Chapter 10, "Habitat Around Your House.")

A 15-ACRE FIELD

An ideal planting site for a warm-season grassland is a weed-free field from which corn or some other crop was harvested the previous year and where the ground is mostly bare. It will be much easier to successfully establish a grassland on such a site than on an old pasture or a field that's already growing cool-season grasses—but you can certainly turn a pasture or a fallow field into a healthy native grassland; it will just take more effort.

Before launching a project of this size or larger, contact your local USDA Natural Resources Conservation Service office. (NRCS has offices in all states; the phone book or an internet search will get you to your local one.) An NRCS specialist will make a site visit and advise you on which grasses and cultivars will do best on your site, plus draw up a plan for site preparation and planting. In some cases, an evaluation of your site may reveal that warm-season grasses are present but are being suppressed by other plants. In that case, you may be able to conduct a controlled burn, remove competing trees and brush, mow, or apply herbicides to restore the warm-season grasses without needing to replant.

NRCS has cost-sharing programs for creating and improving grasslands. They may encourage you to enroll in the Conservation Reserve Enhancement Program (CREP), letting you receive rental payments for taking environmentally sensitive lands out of cultivation to help conserve key wildlife and plant species. CREP enrollments are voluntary and generally last 10 to 15 years.

Your plan may direct you to plant taller warm-season grasses—big bluestem, switchgrass, and Indian grass, which grow to six or seven feet—or mix in some shorter grasses such as little bluestem. The plan may include areas of cool-season grasses, perhaps as a border around a field of warm-season grasses, to act as a fire break should the warm-season grasses catch fire. A plot of cool-season grasses will also offer some food and cover in early spring, before warm-season grasses have begun to grow, and later in the year after the warm-season grasses have become dormant. NRCS can offer advice on the best wildflowers to mix in with warm-season grasses to help native pollinators, including bees, flies, butterflies, and moths.

Site preparation may call for cutting existing growth; disking to chop up weeds or crop remnants and give access to the soil; and applying herbicide to lessen competition from unwanted plants. Big bluestem, little bluestem, and Indian grass have airy, fluffy seeds that should be planted with a piece of machinery called a no-till native grass drill. Your NRCS consultant or county extension agent may know farmers you can hire to prepare and plant your plot. A local chapter of a wildlife group such as Quail Forever or Pheasants Forever may have a native grass drill that can be borrowed or rented. Switchgrass seeds can be planted with a conventional grass drill.

NRCS can probably line you up with a contractor (sometimes they're called "practitioners") to do the seeding. Contractors may supply seeds and the equipment for preparing a site, planting, and bedding in the seeds using a tractor-drawn roller. John Smith, in the paper "Managing Land Ecologically Using Perspective of Scale," writes that preparation and custom planting should cost "in the neighborhood of $4,000 to $5,000" for a 15-acre field. Funds from state or federal cost-sharing programs can reduce that expense.

During their first year, warm-season grasses develop deep root systems, and above-ground growth can be hard to see. Some seeds may remain dormant until the second growing season comes around. As grasses are growing, you may need to stave off weeds and invasive plants by spot-mowing, brush-hogging, or applying herbicides. Early on, while the young warm-season grass plants are still short, you can control weeds by mowing the plot down to a height of six to 10 inches. You may need to mow a few times, whacking down the weeds until the grasses have grown above them. When the grasses are covering most of the site, you shouldn't need to mow again for a few years.

Once grasses are fully established, you may not need to do much maintenance at all, since they will exclude and outcompete most weeds. Some grasslands become very thick over time and may need mowing every four or so years. Haying and grazing may also be options if grasses have been planted for combined agricultural and wildlife purposes.

A landowner can also use one more tool to keep a grassland healthy and vigorous: fire.

PRESCRIBED BURNING

Warm-season grasses evolved in environments that had periodic fires caused by lightning, Native Americans' management practices, or both. The grasses' deep roots let them grow back quickly following a fire. Fire also returns nutrients to the soil, which promotes a flush of new growth.

Using fire to manage a grassland or any other wildlife habitat is known as "prescribed burning" or "controlled burning." The term "prescribed" refers to burning as part of a prescription, or plan, to achieve a desired end. Conservationists and landowners use fire to lessen the amount of flammable matter on the ground, preventing hotter, out-of-control fires in the future; to limit the growth of unwanted trees, shrubs, or low plants, both native and invasive; to keep grass plants healthy; and to help certain fire-adapted species, such as longleaf pine, blazing stars, wild lupine, and sandplain gerardia, the seeds of which need low-intensity fire to germinate.

Spring is the best time to burn warm-season grasses, which grow back quickly once the weather warms up. In a grasslands, a fire will burn off the thatch (accumulated dead vegetation on the ground) that can gradually make a habitat less useful to birds and other animals that walk or run on the ground through grasslands to find food or escape from predators. Burning a field may destroy birds' nests, but in general, grasslands burned every three to four years will have higher bird-nesting densities than ones that are mowed instead of burned. To set back trees and shrubs, conservationists conduct burns in summer after the trees and shrubs have put forth leaves so that burning produces the maximum stress in those target species. Burns in autumn encourage the growth of forbs (herbaceous flowering plants).

It would be dangerous, irresponsible, and probably illegal for a private landowner to simply wait for a dry spell and set a field of grass on fire. Fortunately, habitat-management specialists can be hired to plan and safely carry out burns. You can find certified contractors by contacting your local NRCS office, your state wildlife or forestry agency, or your state's Prescribed Fire Council. (Not all eastern states have fire councils; try an internet search using the terms "prescribed fire" along with your state's name.) NRCS may have funding available to defray the costs of prescribed burning. The Nature Conservancy (TNC) has a long history of using fire to help restore natural ecosystems on their properties; TNC is an excellent source of information about prescribed burning and may offer advice on how to fund burns to improve habitat for wildlife or plants. Contact your state TNC office to learn more.

Trained fire specialists can draw up comprehensive plans describing management objectives

Prescribed burning can lessen the amount of flammable matter on the ground and limit the growth of unwanted trees and shrubs in grasslands. Spring is the best time to burn warm-season grasses, which will grow back quickly once the weather warms up.
Joel Carlson

(such as making a grass field a better habitat for bobolinks and meadowlarks), site preparation, safety precautions, needed weather conditions, adherence to local fire or smoke regulations, and a step-by-step procedure for how a fire will be set and contained. To conduct burns, contractors put together crews of men and women familiar with local weather conditions, fuels, fire behavior, laws and regulations, and nearby communities. It can take up to a year to plan, organize, and successfully carry out a prescribed burn.

Joel Carlson is a fire ecologist and principal consultant and owner of Northeast Forest and Fire Management, a Massachusetts company that conducts 25 to 30 prescribed burns a year for wildlife organizations, land trusts, state and federal agencies, towns and municipalities, and private landowners. He calls prescribed fire "a tremendous technique for managing habitat for wildlife."

Carlson is certified to write prescribed burn plans and conduct burns in New England and New York. Silvix Forestry, run by Shannon Henry, offers similar services in Pennsylvania. Farther south and in the Great Lakes states, many private foresters are certified by their respective states to plan and carry out burns.

According to Carlson, "The larger the burn, the more cost-effective it will be." Burning in highly developed parts of the Northeast—such as the area around Carlson's home on Upper Cape Cod—is more expensive than burning in a rural part of the South, for example, where there may already be a strong tradition of using fire to manage farms and forests and to improve wildlife habitat. In the Northeast, the greater number of smaller landholdings, combined with proximity to roads, towns, and schools (Carlson calls it "the urban interface"), can make prescribed burning extremely challenging. In the South, burning a large field may cost as little as $5 per acre. In the Northeast, burning a 15-acre field of warm-season grasses may cost $333 to $800 per acre, for a total of $5,000 to $12,000, with a probable average of around $7,000. (While that sounds expensive, it's actually less costly than some other commonly used management techniques, such as noncommercial timber harvests, in which low- or zero-value trees are cut down to create wildlife habitat, with no resulting income to defray expenses.)

"On large burns of 200 acres or more, we've gotten costs down to around $40 per acre," Carlson says. He conducts burns on a variety of habitat types, from warm-season grass fields to the pitch pine and scrub oak forests that are common on Cape Cod, where he has done a number of burns to create habitat for the New England cottontail, that region's native rabbit. "In areas with many houses, towns, and roads, we have to be very careful so that fires don't burn out of control," Carlson says, "and so that smoke doesn't carry into residential areas or cause traffic accidents by reducing visibility on roads." On some projects, "we do mechanical treatments first—basically use a machine to shred some of the standing vegetation—to ensure a safer, less-intense burn that still renews the habitat."

If you are considering a habitat project that might include prescribed burning, Carlson recommends that you consult with a fire professional early in the planning process. Let's say you want to plant a 15-acre grassland as one component in a habitat to help bobwhite quail, American woodcock, box turtles, and pollinating insects such as native bees, butterflies, and moths. A fire professional might suggest you put in a 10- to 12-foot border of cool-season grasses all the way around the plot, creating a fire-break that will be effective no matter which direction the wind is blowing on the day of a prescribed burn. He or she might suggest that several burns take place over a span of years to keep the grassland healthy—or to restore a neglected grassland or other habitat. "Rarely is a single prescribed fire adequate to meet habitat goals and objectives," Carlson says. "Often multiple burns are required, perhaps with alternative treatments to make the burning safer or more effective."

A certified contractor can develop a prescribed burn plan, a legal document that will protect the landowner, the prescribed-burn contractor, and the burn crew. Such a plan might cost a landowner $2,500 to $4,000, an expense that could be lessened through cost-sharing by a state or federal agency. Like a timber-cutting plan that's integral to a larger forest management plan, a prescribed burn plan might direct that burning take place three times over the course of 10 years. After 10 years, a burn plan can usually be updated for a lower cost.

Carlson has run Northeast Forest and Fire Management for 12 years. During that time, he says, attitudes toward fire have changed. In the past, many people thought fire was a destructive force that needed to be kept out of woods, shrublands, and grasslands at all costs. Says Carlson: "Today, more and more state wildlife and forestry agencies throughout the East are adopting prescribed burning as a valuable technique to reduce the flammability of the landscape and to make and maintain a variety of habitats for wildlife and plants."

That new understanding has spread to other groups as well, particularly land trusts wanting to re-establish natural cycles and habitats. Education about the true effects of fire and prescribed burning is key. "I think the same kind of evolution is taking place in landowners' attitudes," Carlson says. "As more and more people find out how useful—and natural—prescribed burning can be, I predict they will embrace this safe and effective tool."

6

Shrublands for Wildlife

A TECHNICAL MANUAL FOR MANAGING eastern habitats defines shrublands as "areas of woody plants typically less than 10 feet tall, with scattered open patches of grasses and forbs between them." In any particular shrubland, "grasses and forbs" could include planted cool-season grasses, native warm-season grasses, and weeds and wildflowers. Sedges grow between shrubs on some sites.

Powerline rights-of-way and abandoned farm fields have such habitat. Shrublands also include alders growing on the damp ground in and near wetlands; stands of mountain laurel, an evergreen shrub that grows on acidic soils in forests, mainly in mountainous regions; and scrub oak, a bushy, low-growing oak found on sandy barrens, rocky hillsides, and mountain plateaus. Old beaver

Shrubland habitats have dwindled because of development, invasion by exotic plants, and forest succession, when the growth of trees gradually turns a shrub area into woods. More than 60 kinds of shrubland wildlife have been classified as "species of greatest conservation need" by eastern states.
Kelly Boland

flowages may pass through a shrub period on their way back to being forest. Some conservationists add recently clearcut woods to the "shrublands" category, especially when shrubs in the forest understory are newly invigorated by sunlight after shade-producing canopy trees have been removed by logging.

Shrublands are a dwindling habitat: Over the last 40 to 50 years, many thousands of acres have been lost to development and to forest succession, the natural process through which the growth of trees gradually turns a shrubby area into a woodland. Many kinds of wildlife require shrublands or young forests (more than 60 are listed as "species of greatest conservation need" by eastern states), and many other more common animals frequent them. Shrubs produce fruits and nuts that are a mainstay in many animals' diets. They also offer food in the form of buds, catkins (pendant flowering structures), and seeds. Insects, a prime food for many amphibians, reptiles, birds, and mammals, are often abundant in shrublands. Shrublands attract wildlife because of the way they combine excellent food and good hiding cover in the same place.

Many birds nest in shrublands. They feed their nestlings with high-protein insects that they catch among the leaves of native shrubs.
© *iStock.com/photographybyJHWilliams*

You can create a shrubland by letting a field grow up with shrubs. You can speed up the process by planting shrubs, though you may have to protect them from deer. Shrublands can sometimes be brought into existence by clearcutting stands of timber with shrub understories and then coming back every couple of years and cutting down the trees' regrowing shoots and seedlings.

It's easier to maintain and improve an existing shrubland than to create one from scratch. The kinds of plants in such settings—and the long-term stability of the habitat—depend on many factors, including what was growing in an open area in the past, how many years have gone by since a field was taken out of farming, local sources of seeds for shrubs (including both native and invasive types), soil quality and moisture, whether the site is flat or sloping and the direction in which any slope faces, a history of fire, and local climate.

Thick stands of shrubs suggest that healthy, extensive root networks lay beneath a field while it was being grazed, mowed for hay, or periodically burned by fire. When those disturbances no longer kept the above-ground growth pruned back, shrubs shot up from the mature root systems already in place.

Some shrublands can be fairly stable, with shrubs persisting on a site for up to 40 years with little or no help from humans. Once they've been established, dense shrub clones can exclude invading trees for years. However, if the patch already has abundant trees, its life will be shorter—on the order of 10 to 20 years—if the trees aren't removed.

Native shrubs that commonly come up in old fields include juniper, pasture rose, dogwoods (flowering, pagoda, gray, silky, and red osier), blackberry, black and red raspberry, elderberry, beaked hazelnut, viburnums (arrowwood, nannyberry, black haw, and highbush cranberry), buttonbush (especially in wet settings), black and red chokeberry, witch hazel, spicebush, ninebark, hawthorns, chokecherry, pin cherry, sumacs, and alder. Vines in shrublands can include Virginia creeper, greenbrier, and poison ivy. Small- to medium-sized trees that commonly grow in old-field shrublands are gray birch, quaking aspen, eastern red cedar, downy serviceberry, American holly, persimmon, Osage orange, redbud, wax myrtle, sassafras, crabapples, and black locust. Larger trees can include ashes, maples, oaks, tulip tree, sweetgum, black gum, and black cherry. Although not native to North America, apple trees, an important food source for wildlife, often grow in shrublands and old fields. In the South, native conifers that invade old fields are Virginia, pitch, longleaf, slash, and loblolly pine. In the North, white pine, tamarack, and jack pine may show up. White spruce can establish itself in an old field; the tree's winged seeds blow in on the wind and work their way down into the sod, and the resulting seedlings soon rise above the grasses, wildflowers, and shrubs.

In old fields with many woody shrubs and herbaceous plants, wild animals take advantage of the structural diversity (vegetation of different heights) to find nesting sites and hiding cover. Cottontails and other small mammals duck into shrub thickets to evade foxes, bobcats, weasels, hawks, and owls. American woodcock nest on the ground in thick stands of alder and beneath other small trees and shrubs, and they also probe for worms (their primary food) in those places. Blue-winged warblers, golden-winged warblers, brown thrashers, mockingbirds, and gray catbirds are a few of the many songbirds that nest in shrub habitats. Various birds also eat the insects whose life cycles include native shrubs. Patch size can be a factor: Small, isolated shrublands aren't big enough for New England cottontails, yellow-breasted chats, and golden-winged warblers, which all need shrub patches of at least 25 acres or larger. In general, the larger the shrubland, the better it will be for local wildlife. However, even if you have only an acre of shrubland, you are the steward of a high-value wildlife habitat that is definitely worth preserving and managing.

You can often improve the wildlife value of a shrubland by increasing native plant species diversity and structure. Ten to 30 percent of an old field covered with shrubs and young trees is a good minimum. You can plant clumps of different species of shrubs, fencing them against damage caused by deer. Often a four-foot fence will be high enough to divert those indefatigable browsers.

Native crabapples are shrubs and small trees that you can plant or encourage. Birds nest among their branches; bees and other insects pollinate their showy, fragrant flowers; and many animals eat their fruits in late summer and fall. The American or sweet crabapple (*Malus coronaria*) occurs in central and eastern North America. Prairie or Iowa crabapple (*Malus ioensis*) is native to the Midwestern states. Southerners can plant Southern crabapple (*Malus angustifolia*), which forms thickets 12- to 25-feet tall and wide, sending up shoots from its roots as they spread underground. Southern crabapple grows from Delaware south to Florida and west to Texas, often at low elevations. Bobwhite quail, ruffed grouse, squirrels, opossums, raccoons, skunks, foxes, bears, and deer are among the animals that feed on crabapples.

Crabapples and many other beneficial and productive native shrubs can be obtained from

WILDLIFE SKETCH

Wave the Foliar Flag

Fruits of flowering dogwood are rich in fat—perfect for songbirds that must replenish their energy reserves during the autumn migration. Brightly colored leaves and fruits attract the birds, which disperse dogwood seeds in their droppings.

HAVE YOU EVER WONDERED WHY THE LEAVES OF SHRUBS AND VINES and small trees such as dogwood, nannyberry, Virginia creeper, sassafras, and black gum turn such brilliant colors in early autumn? Surely not to please the sensibilities of hikers and grouse hunters. No, there's a more urgent biological reason behind this luminous display: to catch the eyes of migrating birds and get them to eat the plants' fruits.

Biologists call this type of advertising "foliar fruit flagging." Take flowering dogwood as an example. The leaves of this large shrub or smallish tree turn many shades of red, from rich burgundy to stop-sign scarlet. Birds see the gaudy display and pause. Dogwood berries, called drupes, seem bitter and unpalatable to humans, but the birds don't mind. The berries (eye-catchers themselves, with their own sizzling scarlet coloration) have a high-fat content, up to 24 percent by weight. For a bird, a meal of dogwood berries becomes important energy to keep warm on chilly nights and to fuel further flight southward. Ripe dogwood berries don't stay on the shelf in nature's pantry for long. Flickers, vireos, robins, thrushes, bluebirds, catbirds, cedar waxwings, and a host of other migrating birds, both long- and short-range, readily eat these high-value foodstuffs.

Sassafras's mitten-shaped leaves turn dazzling colors, from yellow to orange to red and every shade in-between. Their blue, berrylike fruits sit cupped at the tops of crimson stalks. Nannyberry, a viburnum, lures with showy orange-red foliage and bluish-black fruits on slender red stalks. Down at ground level and climbing up tree trunks and fence posts, Virginia creeper entices with spectacular red leaves and contrasting blue-black, grapelike berries. The dark-colored fruits might be easy for a flying bird to miss—if not for those emphatic foliage colors that practically shout "Free food! Come and get it!"

The fruits mentioned above have a thin outer skin, pulpy flesh, and a stony center protecting a hard, dense seed. The fleshy part of the package is what the birds are after; the seed comes along for the ride. After the birds gobble down the fruits, the seeds pass through the birds' digestive systems, "scarifying" them, weakening the seeds' outer coats and making it easier for them to germinate the following spring. The birds deposit the seeds in their droppings, generally some distance away from the plant that produced them, letting the parent plant increase its range and ensure its perpetuation.

The leaves of staghorn sumac, chokecherry, arrowwood, and poison ivy are often spectacular, but the fruits of those plants don't get snapped up as readily by birds—which seem to look past the overhyped advertising, knowing that the fruits don't have quite as high of a fat content. Such lower-value fruits hang on their stems for a longer time, even into the depths of winter—so much the better for the resident birds that tough it out and overwinter in the eastern woods.

In areas where deer populations are high, a four-foot fence can divert these browsers, letting planted shrubs become established. Planting clumps or strips of shrubs in old fields boost their value to wildlife.

commercial nurseries, state wildlife agencies, local conservation districts, and online sellers.

Over time, through selective brush-hogging, you can let some areas in your shrubland become denser than others; that way you'll end up with a mosaic of shrubs of different sizes and heights throughout the patch. Another option is to add more low structure by clearcutting the trees on an acre or two of adjoining woods. In the same vein, if your shrubland is surrounded by forest, you can create a border of fresh thick growth all around it by cutting down a band of trees 50- to 150-feet wide in a scalloped fashion, with the border extending further into the woods in some places than in others. You can sell the trees, mill them into lumber, use them for firewood, or simply leave them where they fall so that their tops and branches form a natural hedge against deer browsing. Use some of the trees' trunks and branches to build brush piles throughout this border of young, regrowing forest to make it an even better habitat.

Building Brush Piles

Brush piles are easy to build and greatly increase the cover quality of many different habitats, including shrublands, wildlife corridors, and the young forest that springs up following a timber harvest. Wildlife uses brush piles for nesting, resting, evading predators, keeping cool during summer, and staying warm and dry when it's cold and stormy. (In winter, the inside of a brush pile can be several degrees warmer than the air outside.) Rabbits home in on brush piles, which also get used by songbirds, turtles, lizards, snakes, opossums, shrews, weasels, chipmunks, squirrels, woodchucks, and even black bears if the piles are big enough.

Brush piles can be many different sizes, but a good standard dimension is 20 feet across and four- to eight-feet tall. Start with a foundation of logs six to 12 inches in diameter spaced about a foot apart, making a grid and allowing wildlife easy access at ground level. (You can also use large rocks for the first layer, a combination of rocks and logs, or cement blocks.) Lay the second layer of smaller logs on top of and roughly perpendicular to the base layer. Keep building upward, repeating with progressively smaller logs and branches, then top off the pile with loose brush or pine boughs. Add new material to the top every couple of years to extend the brush pile's life. In linear habitats such as wildlife corridors, build a brush pile every 100 yards, or even more closely together if you have the time and materials. In other habitats, one to three brush piles per acre provide ample hiding cover. (Don't build brush piles on sites where you may want to conduct prescribed burns in the future.)

A small tractor with forks can lift and place the large bottom logs; smaller materials can be layered on by hand. If you're planning a timber harvest, you'll probably have a heavy machine on-site, such as a grapple skidder. Have your forester specify in your logging contract building a certain number of brush piles per acre. A good skidder operator can quickly assemble a brush pile using logs and branches that won't be sold as wood products.

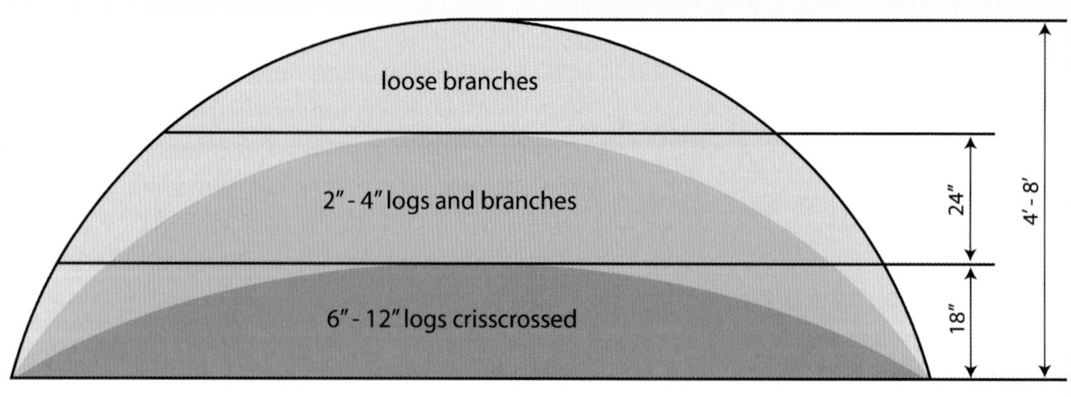

This diagram illustrates one way to build a brush pile. You can also use large rocks or concrete blocks for the bottom layer. Brush piles add cover value to many different types of habitat, including shrublands.
Maja Smith

If your goal is to keep a healthy, permanent shrubland, periodically cut back trees as they invade the site, before they grow too tall. If you already have some fairly large trees in your shrubland, consider leaving a few of them, especially if they offer important food to wildlife: walnut, butternut, cherry, black gum, hickory, and oak would be candidates for preserving, as would shorter trees such as persimmon, American holly, and serviceberry. Trees that yield seeds that are somewhat less important as foodstuffs for wildlife, such as birches, maples, and ashes, can be felled or killed in place (girdled or killed with herbicides), turning them into snags that provide perches for birds, roosting sites for bats (they rest beneath the bark once it starts coming loose), and sources of cavities for birds and small mammals. Smaller trees can be pushed over with a bulldozer or log skidder; they can also be felled with a hinge-cut, in which the trunk is cut most of the way through and the tree pushed over. Putting up nest boxes along the edges of a shrubland provides important nesting and roosting sites, with a strong possibility of attracting bluebirds. A good approach is to place nest boxes on poles with predator guards or baffles to keep out raccoons, rodents, snakes, and cats. The edges of shrublands are also prime places to erect roosting boxes for bats.

When aging shrubs get leggy and open, their value as habitat lessens. Cutting them back—using a chainsaw or a large machine—stimulates their root systems to send up many new shoots.
Ted Kendziora, USFWS

Mowing or mulching older overmature shrubs will cause them to grow back more thickly, sending up many smaller stems from their root systems. Mow older shrubs in winter to get the best regrowth the following spring. Use a brush hog pulled by a tractor or an ATV, a chainsaw, a gas-powered handheld brush cutter, or a pair of loppers if your shrub patch is small. For large tracts, you can hire someone to do the work using a tracked machine with a mulching head. Don't mow or mulch the whole field all at once. Leave some areas unmowed each year so they'll continue providing food and cover during the following several years as you renew different areas of the shrubland. This approach will keep some cover in place, plus provide an ongoing supply of berries and other fruits for songbirds that stop among the shrubs while migrating.

Mowing tips: Mow or brush hog old fields every two to five years, depending on site conditions. Mow around existing shrubs (unless you intend to cut them back to rejuvenate them) and around wet areas and rocks to let shrubs establish themselves in those places. Mow after the primary nesting season for songbirds, which runs as late as mid-August—or mow even later in autumn so that late nesters such as goldfinches can bring off broods and so that butterflies can continue to get nectar from fall flowers. If cool-season grasses grow among your shrubs, set your mower deck at a height of four to six inches. For warm-season grasses, set it at six inches or greater. If you suspect that wood turtles, box turtles, rat snakes, or other reptiles may be using your field, mow after October 1 and set your mower at six inches or higher.

Two other techniques for maintaining a shrubland are prescribed burning and grazing. Burning an old shrubland every several years will help kill encroaching trees and invasive shrubs while stimulating grasses and forbs. Burning in April to early May favors native grasses such as broomsedge and can knock back non-native cool-season grasses such as fescue. A burn during the dormant season for plants, from November through March, favors forbs. Many native shrubs found in old fields are resistant to damage from fire. Once fully established, they can be top-killed by fire but will quickly resprout. If your shrubs are less than 10

LANDOWNER STORY

Teaming Up to Help Woodcock and Trout

IN NATURE, WILDLIFE AND HABITAT COMBINE and synergize in intriguing ways. Set out to help wild trout by planting shrubs and trees to stabilize stream banks and cast cooling shade on the waters, and suddenly you're helping shrub-loving woodcock and songbirds as well.

Doug Bierly, a township supervisor who lives in Penns Valley in central Pennsylvania, admitted that some of his motivation in planting a buffer zone along Upper Penns Creek near his house was "so that I didn't have to keep mowing a horse pasture our family no longer needs." Now a strip of native crabapple, silky dogwood, and speckled alder, plus fast-growing aspen, black locust, and redbud saplings, follows the stream as it snakes across Bierly's land: 500 shrubs and trees planted by the U.S. Fish and Wildlife Service on four acres. "Ever since that buffer strip went in," Bierly said, "my wife and I have seen more and more birds." He promptly listed field sparrows, chipping sparrows, indigo buntings, cedar waxwings, robins, grackles, "even kingfishers, which we never used to see here."

Back in 2012, valley resident Lysle Sherwin—at that time, head of the Center for Watershed Stewardship in the forestry department at nearby Penn State University—was one of a team of conservationists who helped inform Bierly and other Penns Valley residents about the technical advice, planning, and financial help they could get from government agencies and conservation groups whose missions include creating habitat for fish and wildlife. The overall project, spearheaded by a local citizens' group, the Penns Valley Conservation Association, has been gaining momentum—and signing up landowners—ever since.

Sherwin said that landowner-to-landowner networking is essential for this kind of watershed-wide effort to succeed. "Property owners in the area are a mix of longtime residents, including families who have farmed the land for generations, along with more-recent move-ins and retirees who like living in a rural setting. Hunting, fishing, and wildlife-watching are strong local traditions."

The U.S. Fish and Wildlife Service (through its Partners for Fish and Wildlife program), the USDA Natural Resources Conservation Service, Trout Unlimited, the Chesapeake Bay Foundation, the Pennsylvania Game Commission, the Pennsylvania Fish and Boat Commission, and the National Fish and Wildlife Foundation have all lent support. Landowners like Bierly purchase, plant, and maintain tree and shrub seedlings, or pay independent contractors to take care of those tasks, and—depending on the type and scale of project that the landowner opts for—the conservation agencies and wildlife organizations contribute to or even cover the costs.

One Penns Valley resident is Lisa Williams, state woodcock biologist for the Pennsylvania Game Commission. She immediately saw the value to woodcock of planting shrubs and trees along Penns Creek, a nationally renowned trout stream, plus

Landowner Doug Bierly, right, studies a map with Lysle Sherwin, a conservation consultant who has helped enlist many property owners to create habitat in the Penns Creek watershed in central Pennsylvania. Bierly planted shrubs and trees for food and cover and so that their shade would cool stream waters.

several smaller streams that feed into it. "Woodcock will feed and rear their young in the thickets that will grow to shade the waterways," she said. "They'll use nearby more-open habitats for spring courtship and nighttime roosting in late summer." In spring 2011, before the watershed project got underway, Williams ran a woodcock singing-ground survey in the area and heard several males along a two-mile route. "That told us a breeding population was already in place," she said, "one that would fill any new habitat that landowners created."

Sherwin and the Penns Valley Conservation Association reached out to residents through letters, brochures, habitat site tours, and a panel discussion featuring representatives of the participating conservation entities.

Landowners enrolled for a variety of reasons. Some wanted to prevent invasive shrubs from overrunning natural areas. Others thought it made sense to turn pastures that were no longer being grazed by livestock into wildlife habitat. Still others wished to improve opportunities for hunting deer, bear, wild turkeys, and ruffed grouse, all of which will use the new young forest and shrubland habitats for nesting, feeding, sheltering, and as movement corridors.

The Bzdil family has done many habitat projects on their land, which includes sections of Pine Creek and Elk Creek in the Penns Valley watershed. Most of the projects were begun by John Bzdil, an orchardist near Sunbury, a town along the Susquehanna River about 50 miles east of Penns Valley. John died in 2013, but his generous attitude toward helping wildlife is carried on by his wife, Theresa, and his two sons, John and Mike. "We started this whole effort 30 years ago," Theresa said, "when we bought a hunting camp surrounded by 60 acres." Now the family owns more than 400 acres, including more than a mile and a half of three streams that support wild native brook trout. "A lot of people want to make improvements to help wildlife, but they don't know how to get started," Theresa said. "They need somebody to give them the right information and walk them through the process a little bit."

That's exactly what Lysle Sherwin and the Penns Valley Conservation Association have been doing. Sherwin retired from Penn State in 2013, but he continues his work as a member of PVCA and through his consulting firm, Seven Willows LLC, based in his home beside Penns Creek near the town of Spring Mills.

Over the years, the Bzdils' projects have included building two wetlands on designated Wetland Reserve easements on 42 acres, plus putting up more than 20 wood duck nesting boxes—"Wood ducks were my husband's favorite bird," Theresa said. They've also erected bluebird and barn owl nesting boxes and planted hundreds of native trees and shrubs.

The Bzdils' son John said, "My father always told my brother and me that we're merely stewards of the land, and of this property, for our lifetimes, so we should work to conserve it and keep it healthy. He taught us that all wildlife, from the smallest insect or bird up to the biggest black bear, plays important roles in nature. My brother and I are already getting our children involved. They help with planting trees. They have lots of questions about nature," John added with a grin, "and we love to answer them."

He listed some of the wildlife they've seen: "Deer, of course, plus bear, bobcat, coyote, fisher, eagle, otter, waterfowl and herons on the constructed wetlands, lots of songbirds, including wood thrushes, and grouse, turkey, and woodcock. We hunt, but it's not the driving force for us. We consider ourselves conservationists first and hunters second."

Rebecca Bragg is director of operations for the Penns Valley Conservation Association. "People are visual learners," she said. "If they can look at their neighbors' properties, over days and weeks and years, and see the real impact that habitat creation has on wildlife, they will be a lot more likely to do a similar project themselves. Sometimes when you read about wildlife habitat and stream restoration, what you hear is a bunch of complicated scientific terms. But when folks see the actual habitat work, it all becomes more tangible, accessible, and possible."

According to Lysle Sherwin, as of spring 2017 the watershed project had attracted 24 landowners with more than 1,800 acres of land, in parcels ranging from four acres up to several hundred acres in size. The watershed work got a huge boost in 2018 when the National Fish and Wildlife Foundation gave PVCA more than $100,000 (the grant will be matched by funding from the Pennsylvania Department of Environmental Protection and the U.S. Fish and Wildlife Service) to create 20 acres of young forest and shrubland habitat to shade and improve 3,500 feet along streams that have brook trout.

Even though the Penns Valley work centers on one watershed, its partners hope it will have a broader impact. "We realized that if we could figure out the critical pieces in creating a watershed-wide effort like this one," Sherwin said, "the process could be used in other areas." He hopes that local conservation groups throughout Pennsylvania and the East will pick up on this approach: putting landowners in touch with conservation agencies and organizations, and then providing consistent help and guidance as projects are planned and develop, so that they succeed in creating or improving the habitat that fish and wildlife need.

As Lisa Williams, the state woodcock biologist, has continued to run springtime singing ground surveys in Penns Valley, she has heard more male woodcock in areas where landowners have made and improved habitat. She believes the valley will become increasingly important for woodcock as other landowners learn about the cooperative effort and join in. "Our watershed will produce more local woodcock, and migrating woodcock and other birds will stop and rest in places where they'll find food combined with cover to protect them from weather and predators. Just think what could happen if projects like this were replicated up and down the Atlantic Flyway," which is one of the major flight corridors that woodcock and songbirds follow as they migrate between northern breeding grounds and southern wintering areas.

When many landowners work on different habitat projects in a local area, their combined efforts multiply the benefits to the environment—and also to human residents, whose enjoyment of the outdoors increases as they see more fish and wildlife. The Penns Valley project is also increasing folks' awareness of the need to permanently protect land for wildlife; lately, PVCA has been referring property owners to three land conservancies and a county farmland preservation program, all of which are acquiring easements and development rights in the area.

"People want to do the right thing for wildlife," Sherwin said. "If we as conservation professionals help them find the best and easiest ways to do it, good things can and will happen."

years old, fire could damage them badly or kill them; you can protect them from fire by disking or mowing around the thickets.

Goats, some sheep, and a few cattle breeds, such as Scottish Highland and American Milking Devon, will browse both herbaceous and woody plants and can be used to keep small trees from invading a shrubland. Contact a livestock specialist with your university cooperative extension service or the USDA Natural Resources Conservation Service for advice on setting up a grazing plan.

WET SHRUBLANDS

Alder, silky dogwood, red-osier dogwood, willows, buttonbush, high bush cranberry, winterberry, sweet pepperbush, and highbush blueberry grow in soils edging streams and ponds (including beaver ponds) and in wet swales and other low, damp areas. These shrubs can be planted using either bare-root or containerized stock. Shrub willows and red-osier dogwood grow well from "live stakes": sections of upright branches without twigs or leaves that are cut from winter-dormant plants, placed in water, and stuck or pounded into the soft soil within one day. (Live stakes can also be kept in cold storage for planting later.) Stakes that are three-quarter inch to one and a half inches in diameter and two to five feet long can be handled easily. Search online to learn the ins and outs of planting live stakes to create shrub habitat.

Wet shrublands often persist naturally, sometimes for decades, as wet soils combine with periodic flooding in spring and fall to keep trees from taking over. The time to manage a wet shrubland is when the shrubs are no longer dense and start losing their vertical structure. Alders, for example, become "leggy" and straggling when they get old, with only a few thick stems that tend to grow horizontally instead of vertically; sometimes the stems break off and tip over. Grasses and sedges can grow too thickly in the partial shade beneath old alders, making a site less useful for woodcock and some other kinds of wildlife that need access to the bare or mostly bare ground.

You can cut down competing trees such as red maples, black willows, cottonwoods, and tamaracks if they become too abundant and start casting shade on shrubs. In summer, a tree's roots take in water that is transpired through the leaves, helping to keep the tree cool; transpiration by too many trees can dry out the soil in a wet shrubland, changing the habitat.

If you own a large tract of wet shrubland, carefully map it, dividing the habitat into several management blocks. You can hire a contractor with a Brontosaurus, Hydro-Ax, or similar machine to go in during winter when the ground is frozen and cut down or mulch overmature shrubs. They will grow back densely the following spring, sending up the many small shoots and healthy foliage that attract wildlife. Cut your management blocks on a rotation: Don't renew all of the shrubs at once, but instead, cut different blocks a few years apart. If your wet-shrub patch covers only a few acres, you can cut back shrubs simply and cheaply with a chainsaw.

MOUNTAIN LAUREL

Mountain laurel is a forest shrub with springy stems and leathery evergreen leaves; in late spring, it puts forth spectacular pink blossoms. A member of the heath family, it grows on thin, acidic soils in hardwood forests, mainly in the mountains, from southern Maine to Georgia and northern Florida. Other shrubs that grow along with mountain laurel include rhododendron, blueberry, huckleberry, and azalea, which are also heaths.

Usually, when people mention mountain laurel, it's "How can I get rid of it?" since this shrub can definitely impede humans and human activities. Duane Diefenbach, a Penn State wildlife professor who researches white-tailed deer, writes the following: "Anyone who has tried to walk through a stand of [mountain laurel] knows it's faster to walk the half a mile around it than the tenth of a mile through it." Some foresters don't like mountain laurel, as the shade it casts, combined with its dense root system, can prevent the growth of seedlings of trees that ultimately would yield valuable timber, especially oaks.

Mountain laurel provides a modest amount of food for wildlife: Ruffed grouse occasionally eat its leaves and buds, deer browse on foliage and twigs, and some birds eat the shrub's abundant small seeds. Ruby-throated hummingbirds take nectar from the flowers. Mountain laurel's greatest value to wildlife is its ability to produce cover. Forest birds that nest among the shrub's dense stems include common yellowthroat, black-throated blue warbler, chestnut-sided warbler, Canada warbler, worm-eating warbler, hooded warbler, American redstart, and eastern towhee. Towhees and thrushes feed on the ground beneath mountain laurel, as do voles, mice, and other small mammals, whose presence attracts timber rattlesnakes, foxes, bobcats, and coyotes. New England cottontails and Appalachian cottontails use mountain laurel for shelter during winter and to evade predators. Deer often hide out in thick mountain laurel stands.

Like many shrubs, mountain laurel can become thin and leggy when it gets old. Fortunately, mountain laurel sprouts prolifically from its stumps if it is cut down and from its roots if the shrubs are damaged. You can expand a mountain laurel thicket by using the blade of a log skidder or a small bulldozer to scrape it up along its edges. Dragging a log through a patch will cause enough damage so that the shrubs will send up shoots at ground level; mountain laurel seeds also may sprout on bare mineral soil dug up by the log's passage. Cutting off stems with a chainsaw or heavy-duty loppers will also cause the shrubs to grow back thickly. If you are worried about doing permanent harm, try to invigorate or expand one small patch and see what happens. Mountain laurel thins out and becomes less healthy in heavy shade, so cutting a few trees to give the shrubs some extra light may revitalize them.

Large stands of mountain laurel can be a fire hazard because the shrubs' waxy leaves burn rapidly and at a high temperature—something to keep in mind if you have a house in the woods.

SCRUB OAK

If you own or manage extensive acreages in mountain or coastal settings, you may have some scrub oak shrubland. This fascinating ecosystem is centered on dry, infertile, acidic soils, and it sustains rare and uncommon insects as well as larger wildlife. Some hunting camps in upland areas of the Appalachians have such habitats, as do municipalities and large landowners in coastal sand-plain regions.

Scrub oak (also known as bear oak) is a shrub or small tree that forms extensive and sometimes almost impenetrable thickets on rocky hillsides, sandy barrens, and mountain plateaus. Its range extends from Maine through southern New England and New York, south to North Carolina. Most scrub oaks never get taller than four to eight feet, with stem diameters of one to three inches. They may cloak areas that have been swept by fire or repeatedly logged. Other plants with which they're found include huckleberry, lowbush blueberry, bearberry, lupine, New Jersey tea, sweet fern, and golden heather. In some settings, native sedges and prairie grasses such as big and little bluestem grow in openings between the shrubs. Trees in fire-evolved scrub oak communities may also include pitch pine, dwarf chinquapin oak, chestnut oak, quaking aspen, gray birch, black cherry, and pin cherry.

Look for insects such as buck moths, rare black-and-white moths whose caterpillar larvae feed on scrub oaks' leathery leaves. The Karner blue butterfly, which is on the federal Endangered Species list, lives on a few key sites that support native lupine, such as Albany Pine Bush

Scrub oak grows in sandy and infertile soils in coastal and mountainous areas. Many kinds of wildlife eat scrub oak acorns, and rare insects feed on the shrubs' leathery leaves. Selective logging and prescribed burning can keep scrub oak habitats healthy.

Preserve in upstate New York. Eastern towhees, pine warblers, prairie warblers, whippoorwills, common nighthawk, and ruffed grouse are some of the birds in these hardscrabble settings. Grouse, turkeys, blue jays, deer, bear, squirrels, chipmunks, and mice are a few of the animals that eat scrub oak acorns. New England and Appalachian cottontails, box turtles, spadefoot toads, and hognose snakes live in pitch pine and scrub oak communities. In the Appalachians, scrub oak stands provide important habitat for timber rattlesnakes, whose numbers have fallen drastically in recent decades, leading many states to classify this reptile as "threatened" or "endangered."

Scrub oak habitats have been degraded by development and fragmentation on Cape Cod and in other coastal areas and by trees growing tall and shading out the shrubs, especially in places where fires have been suppressed. Logging can remove taller trees. In scrub oak habitats that haven't seen fire for 20 or more years, big machines can mow, masticate, or shred the overmature shrubs during the growing season, followed by controlled burning a year later. Afterward, the shrubs will grow back more densely and offer better food and cover.

If you're lucky enough to own an extensive stand of scrub oak, contact your state's wildlife or forestry agency to learn about options for keeping this uncommon habitat healthy and productive. State field offices of The Nature Conservancy are excellent sources of information on how to protect and manage rare habitats including scrub oak barrens. TNC is probably the foremost user of prescribed burning to repair and restore habitats that need periodic fire.

INVASIVE SHRUBS

It's a sobering fact that many shrublands in the East are badly infested with non-native invasive shrubs. Even in northern Vermont, where I live—a cold and fairly remote place—every year I find myself doing battle with exotic honeysuckles in the pair of two-acre meadows that I manage as mixed old-field and shrubland habitat.

Landowners should do whatever they can to fight invasive plants in shrublands—and also in grasslands and forests because invasives can quickly jump into shrublands from those two other habitat types. It can be difficult, if not impossible, to eradicate all invasives in a shrubland, but if you don't try to control them, they will overrun your habitat. Quite simply, doing nothing about invasives is not an option if you want to keep a shrubland functioning as a healthy home for wildlife.

Some of the invasives that degrade shrublands include mile-a-minute weed, privet, Japanese barberry, Russian olive, autumn olive, burning bush, bush clover, shrub honeysuckles, multiflora rose, Oriental bittersweet, glossy and common buckthorn, and kudzu. Phragmites, bamboo, Japanese knotweed, and European black alder invade wet shrublands. (For more information on these and other unwanted exotic plants, see Chapter 9, "A Plague of Invasives.")

7

Wetlands for Wildlife

WETLANDS ARE PLACES WHERE WATER BUILDS UP FASTER THAN IT DRAINS or seeps away. From an ecological standpoint, a tract of land can be considered a wetland if it has three basic characteristics: Water is present on, at, or near the ground's surface for some length of time during the growing season. The site has *hydric* soils, which develop under saturated conditions. And the site is dominated by plants that are adapted to growing in such soils.

Wetlands with grasses, sedges, cattails, and other low plants are generally called *marshes*, while those with shrubs and trees are known as *swamps*. Many other terms and regionalisms describe and define various kinds of eastern wetlands, including bogs, fens, bottomlands, wet meadows, beaver flowages, potholes, sloughs, pocosins, and Carolina and Delmarva bays. Water in a wetland may be deep, as in a deepwater marsh, or it may stand at or below the soil's surface most of the time, as in a bog or fen.

Conservationists built this vernal pool on Geneva Swamp Preserve, owned by the Cleveland Museum of Natural History in northern Ohio. Salamanders, frogs, and other amphibians readily breed in constructed vernal pools; other wildlife uses them, too.
Jeff Herrick

Before European settlers arrived in North America, wetlands covered some 221 million acres of what is now the lower 48 United States. Since then, we've destroyed more than half of those wetlands by filling and draining them. Colonial-era planter William Byrd II, who surveyed and named the Great Dismal Swamp on the border between Virginia and North Carolina, summed up the way many early settlers felt toward wetlands: "A horrible desert, the foul damps ascend without ceasing, corrupt the air and render it unfit for respiration…a miserable morass where nothing can inhabit."

Contrary to Byrd's beliefs, many life-forms can and do inhabit wetlands, from microscopic algae to massive moose. Different wetland types provide homes for diverse communities of wildlife. Over the last half-century, as we have paid increasing attention to scientific studies of wetlands' functions and benefits, our society has gradually come to recognize their great value—not only as habitats for wildlife, but as reservoirs that soak up and slowly release floodwaters, natural water-purification systems, and places of beauty and productivity. Yet wetlands continue to be degraded by pollution and lost to urban, agricultural, and industrial development.

Muskrats live throughout the East. Considered semi-aquatic, these house-cat-sized rodents eat about a third of their body weight daily, feeding on roots and stems of many wetlands plants, with cattails being a favorite.
Steve Arena, USFWS

GETTING TO KNOW WETLANDS WILDLIFE

A host of wildlife lives in wetlands, including insects, mollusks, fish, amphibians, reptiles, birds such as waterfowl, wading birds, and shorebirds, and mammals such as star-nosed moles, beavers, muskrats, and otters. Some animals live their entire lives in wetlands, while others go there at different times for food or cover or to use wetland mosaics as travel corridors. Frequent visitors include frogs, toads, turtles, swallows, flycatchers, waxwings, ruffed grouse, wild turkeys, owls, hawks, ospreys, eagles, bats, snowshoe hares, rabbits (swamp, marsh, New England cottontail, and eastern cottontail), foxes, coyotes, bobcats, white-tailed deer, and black bear. Woodcock probe for worms in damp wetland soils; when woodcock migrate north in spring and get caught by a late snowfall, they home in on wetlands where the ground remains unfrozen and they can find food. Wetland denizens are not exclusively native animals: The nutria, a South American rodent originally brought to Louisiana for fur production, has invaded many wetlands in southern states, where its heavy feeding damages thousands of acres.

Not all plants survive in wet soils. Those that can are said to be able to tolerate "having their feet wet."

Submerged plants are rooted in soils and sediments and grow underwater. Examples include

The Chorus at Oak Pond

A vernal pool can be a noisy—and an endlessly fascinating—place. Frogs and salamanders breed in these temporary wetlands. Here, a wood frog floats in the shallows. In some areas, female wood frogs are larger and redder than males.
© *iStock.com/Wirepec*

IT FILLED THE MARCH NIGHT: A SOUND LIKE STONES CLACKING TOGETHER. The land sloped down. The leaves were wet from days of rain. Mud sucked at my boots. I crept to the edge of Oak Pond, and the sound filled my ears. It came from all over the pond. It came in waves—strengthening, receding, strengthening again. The waves dissolved into cacophony, scores of sound points that bombarded the surrounding land.

I flicked on my light, and the sound stopped.

All across the pond's surface, paired eyes reflected the flashlight's glare. The eyes blazed in shadowy heads that ducked, or turned away, or remained immobile in the water. A tentative *clack* sounded from across the pond. I turned off the light, sat down on a stump, and waited. The chorus began again.

Oak Pond is a vernal pool. The first chorus there each spring belongs to the wood frogs. It's the same with frogs as it is with birds: The males sing to attract females for breeding, and the sounds are distinctive from species to species.

As I listened, a new voice joined the chorus. This call was a two-note affair—*prEEEP*—the second tone rising higher than the first. The new singer kept calling, the sound thin and piercing above the wood frogs' undertone. Others of his kind joined in. The preeping came from the pond's shallows. I switched on the flashlight and played the beam over the withered grass near my feet. After a few moments, I spotted a frog clinging between two stems, his rump barely touching the water. He was an inch long, tan, his body traversed by dark brown streaks, two of which met in a wavery X in the center of his back. That identified him as a spring peeper, a common frog of eastern North America. The peeper took a gulp of air, clamped his mouth shut, sealed his nostrils, and inflated a pearly vocal sac beneath his chin. He called by shuttling the air back and forth between the vocal sac and his lungs, vibrating folds of skin in the floor of his mouth.

I shut off the light and let the sound wash over me. After a while, I noticed a third voice trying to insert itself between all those clacks and preeps. This new sound resembled a thumbnail rasped across the stiff teeth of a pocket comb—a *trreep* or *crreek*, rising in speed and pitch toward the end. The calling came from a patch of plants in the center of the pond. I didn't bother with the flashlight because I had a good idea who was making the noise: a chorus frog, a brown-and-gray creature about the size of a spring peeper.

I sat and rode the waves of sound. The mingled calling seemed to get inside my head, become white noise, almost hurt. Sometimes the voices would dwindle; then a gust of wind would come moaning through the trees, and the singing would spring back to life.

I stayed at Oak Pond for an hour. The calling never let up. I imagined female wood frogs, spring peepers, and chorus frogs crawling toward the pond from all directions, slipping silently into the water for their annual assignation.

I mucked around at the water's edge, feeling the ripples of swimming frogs, hearing startled yelps and loud splashes. The cone of my light found a gathering of spotted salamanders. The salamanders were black, covered with yellow dots, and big—a good eight inches from nose to tail tip. Four of them weaved and twirled around one another under the water in a silent, hypnotic dance.

When I left the pond, all the way back to the house the chorus was in my ears.

bladderworts, pondweeds, water celery, water milfoils, and waterweed.

Emergent plants reach up and out of the water. *Broadleaf emergent plants* include alligator weed, arrow arum, arrowhead, golden club, pickerelweed, and water plantain. (An exotic broadleaf emergent from Eurasia, purple loosestrife, has invaded and badly degraded wet meadows, marshes, and lake margins throughout the East.) *Narrowleaf emergent plants* are bulrushes, bur reed, cattails, grasses, sedges, soft rush, southern blue flag, spike grass, sweet flag, and wild rice.

Floating-leaved herbaceous plants include floating heart, water lilies, water shamrock, water shield, and yellow pond lily. Duckweed and water hyacinth are *free-floating herbaceous plants*.

Jewelweed, skunk cabbage, ferns, jack-in-the-pulpit, lady's slippers, joe-pye weed, wild mint, goldenrods, asters, marsh marigold, and sphagnum moss are some of the many other plants that can thrive in the hydric soils of wetlands.

Skunk cabbage pushes up in low, wet areas. Black bears newly emerged from hibernation sometimes eat the leaves. Wood ducks, ruffed grouse, ring-necked pheasants, and northern bobwhites consume the seeds, but skunk cabbage isn't considered a major food for wildlife.

"Pterodactyls" and the Muck

Waterfowl, wading birds, and songbirds flock to wetlands for the abundant food and cover they produce. Reptiles, amphibians, and mammals frequent these habitats, too. In New York, Michael Patane has created wetlands on his own land and enlisted area property owners to do the same.

NEAR THE TOWN OF CANASTOTA IN UPSTATE NEW YORK, Michael J. Patane has been building wetlands and extolling their value to wildlife for almost three decades. The land in that region is fairly flat and slopes gently downward toward the north, where Cowaselon Creek drains into Lake Oneida, an 80-square-mile body of water northeast of Syracuse.

"When I was growing up, I hunted ducks in the creek and on the lake," Patane says. "At one point, I realized I wasn't seeing as many ducks as I used to. In 1990, I had just bought an old farm that had everything on it but water. I saw an article in *Ducks Unlimited* [the magazine for the waterfowl organization of the same name] on how to build a wetlands." Hoping to attract more ducks, Patane thought he'd give it a try.

"A big ditch had been dug through the farm to drain the land," he says. "Long ago, the Oneida Indians called this area the Great Swamp. It was a huge impenetrable wetlands before it was converted to agriculture." Derived from decomposed organic matter, the muck soil was highly fertile. "Most of the farmers around here were Italian immigrants," says Patane, whose grandparents on both sides of his family came from Sicily. His father was a lawyer in Madison County; in that capacity, he was part of the effort under a 1954 federal statute known as Public Law 566, the "Watershed Protection and Flood Prevention Act," to obtain easements allowing further draining of the land. Over time, Patane says, "this whole area became a gigantic grid of drainage ditches."

Most farm families owned small tracts of one to 20 acres. "They grew a lot of onions—also celery, lettuce, and other truck crops," Patane says. The heyday for that type of farming stretched from the 1920s into the 1950s. However, "the area was ditched and drained too hard." The soil's fertility lessened, its depth decreased, and erosion set in. Farms became less and less profitable, and starting in the 1960s, many were abandoned. "At one time there were 400 farmers in this area," Patane recalls. "Now there are four."

When Patane decided to make wetlands on the tired old farm he'd bought, he sought advice from the U.S. Fish and Wildlife Service. Carl Schwartz, a biologist with the agency, came and looked at the property. Through its Partners for Fish and Wildlife program, the Service had funding available to create wetlands habitat. Patane explained to Schwartz what he had in mind. "I think we can do that," Schwartz said. He lined up a contractor with a bulldozer.

"We built five small earthen dams with little pothole ponds behind them," Patane says. "They filled up with rain. And the ducks started using them." Next, Patane plugged the ditch that had been draining his farm. In essence, he had begun undoing some of the negative effects of draining the land by returning the habitat to its natural state.

Those initial efforts ultimately would lead Patane to restoring 250 acres of wetlands on his and his neighbors' properties; convincing area landowners to create wetlands and protect them through conservation easements; and starting the Great Swamp Conservancy, where people can come see and learn about wildlife that use wetlands.

Today, the Cowaselon Creek watershed attracts and supports so much birdlife that the National Audubon Society has designated it an "Important Bird Area," or IBA, one of 132 in New York and some 2,700 nationwide. (Through the use of an internationally developed set of criteria, IBAs are identified as globally important sites for the conservation of bird species.)

"At first, all of the wetlands I built were fed by rain," Patane says. "Carl and I had to figure out how to get those wetlands to hold water year-round. We learned to make ponds of different types and different depths." A successful businessman and the owner of a landscaping nursery company, Patane continued to buy land in the area and convert portions of it to wetlands. "From 1990 to around 2000, Carl and I built eight ponds on land that I owned," along with another 20 ponds on other people's properties.

Schwartz also put Patane in touch with the USDA Natural Resources Conservation Service. The NRCS grew out of the old Soil Conservation Service—the same federal agency that had advanced the Watershed Flood and Protection Act in the 1950s. Since then, our culture had begun to recognize the value of wetlands: They improve water quality, absorb floodwater, and support a great diversity of life, including microorganisms, plants, invertebrates, and fish and wildlife. People lamented the fact that millions of acres of wetlands had been drained and destroyed across the country. Imbued with this new attitude, NRCS had changed direction: Now they were enrolling landowners in a Wetlands Reserve Program to restore and create wetlands on their properties, including the played-out farms in what NRCS identified as the Canastota Mucklands Focus Area. NRCS would pay a landowner to set old farmland aside as conservation land, then develop wetlands habitat on it by plugging ditches, disabling buried drainage tiles, and constructing small dams to back up water and create impoundments.

Patane continued to build wetlands. "On my own property I acted as my own contractor," he says. "I would kind of look at it as a big landscaping job. On partner projects, I'd come up with a design, and Carl and I would go over it; he'd look at his budget and say 'I have this amount of funding. Can you do it?' Then I'd rent a D-6 bulldozer or a track hoe and go to work."

In 1996, Patane founded the Great Swamp Conservancy, a nonprofit organization, and negotiated the purchase of five acres with a house, a small barn, and several outbuildings. The tract's original owner later donated more land to the Conservancy, which today takes in 150 acres, most of them in the Wetlands Reserve Program. In addition to some 30 acres of wetlands, the Conservancy has grasslands, young forest, and mature forest, all used by different kinds of wildlife. Seven miles of well-groomed hiking trails vein the property, and a 900-foot boardwalk takes visitors through a restored forested wetland.

Conservancy members and volunteers have put up bluebird boxes, wood duck nesting boxes, and bat roosting boxes. There's an office in the remodeled farmhouse and, in the old barn, an educational and outreach center. The Great Swamp Conservancy is a busy place: Events happen the year around, including festivals linked to the spring and fall migrations of waterfowl, plus programs on aspects of natural history from rehabilitating injured black bears to how a property owner can identify and get rid of non-native invasive plants. Camille Warner, an environmental educator on the Conservancy's staff, leads nature walks and develops lesson plans that integrate with school curricula to teach youngsters about the importance of conserving wildlife and the environment.

Working through the Conservancy, Patane continues to promote wetlands conservation to area landowners. So far, more than 80 landowners have signed easements with the NRCS on some 4,500 acres in the Canastota Muckland Focus Area, which includes between 1,500 and 2,000 acres of restored wetlands.

At age 65, Patane claims rather gruffly to be slowing down and getting fed up with "all the paperwork and fundraising" he finds himself doing as the president and chief executive officer of the Conservancy. His wife, Marilyn, nicknamed Rusty, is the Conservancy's cofounder and the person who organizes all of the events and oversees educational efforts. "She does the real work," Patane says.

In spring, ducks such as hooded mergansers rest and feed in open waters of wetlands. Along with wood ducks, buffleheads, and goldeneyes, hooded mergansers nest in tree cavities near ponds and lakes.

He says he gets the most enjoyment out of being on the land and seeing the wildlife that has benefited from his work. He likes watching waterfowl. In early spring, migrating ducks set down on open water in the wetlands—buffleheads, mergansers, black ducks, mallards. Some of the ducks pass through; others stay and breed. "Blue-winged and green-winged teal nest like crazy here," Patane says. "We've got a great volunteer putting up wood duck boxes all over the place and keeping the old boxes in good repair."

Patane loves it when a new species of wildlife shows up. "Great egrets have been coming in over the last six to seven years. I've seen them hunting in the shallow ponds." Bald eagles soar overhead or perch in snags, looking out for fish and other prey in the impoundments. A survey conducted in 2006 by students with the State University of New York's College of Environmental Science and Forestry, in Syracuse, found 213 species of birds in the Canastota Creek watershed, leading to its Important Bird Area designation. "All of the wetlands, plus the upland habitat work, have restored an important part of the flyway that birds use around Oneida Lake," Patane says. "Before, they were skirting this area because there was little food and cover for them."

Great blue herons nest in a forested wetland on the eastern part of the Conservancy's land. "Last year there were 28 nests," Patane says. The birds build their stick nests in trees that died after their roots were drowned under water backed up behind an impoundment. "People come from all over to see our heron rookery," Patane reports. "It's an amazing place in the spring. It's like pterodactyl heaven, like

Great blue herons have established a rookery—a nesting area—in dead trees in a wetland on Great Swamp Conservancy. When predators such as eagles and hawks fly past, the herons sound the alarm by squawking and flapping their wings.
© *iStock.com/JLFCapture*

watching a prehistoric scene—herons squawking, screeching, and standing up tall in those nests that look like they could fall apart at any moment.

"They're elegant birds, but they look odd—you could call them beautifully awkward birds. It's like they belong in another time. It's fun to watch them when an eagle flies past. The adults go crazy! They squawk and slap their wings and jump up and down—like some kind of bizarre blue heron ballet."

Patane admits he didn't create the entire habitat just for his own enjoyment. "It was also to educate the kids, help them learn what wetlands are all about—and to see some of the things that I was fortunate enough to experience when I was young." He likes it when buses filled with schoolchildren arrive for field trips.

"It's great to pull them away from their computers for half a day. Camille, our outdoor educator, takes them on hikes. They see all kinds of birds and animals and hear all sorts of birdcalls and frog calls, all the natural sounds. It gives them experiences that I hope they'll remember for the rest of their lives."

Wetland shrubs include alders, bog rosemary, highbush blueberry, buttonbush, leatherleaf, meadowsweet, shrub willows, red-osier dogwood, sweet pepperbush, silky dogwood, sweet gale, coralberry, elderberry, viburnums, rhododendron, winterberry, and witch hazel.

Some of the many trees found in wetland settings are black gum, hackberry, catalpa, cottonwood, persimmon, river birch, red maple, silver maple, black ash, green ash, balsam fir, eastern hemlock, American holly, tamarack, white-cedars (Atlantic and northern), hornbeams, swamp white oak, water oak, overcup oak, pin oak, and slippery and American elm.

Most wetlands are not static or fixed habitats, and they may change over seasons or years. Water levels can fluctuate, depending on weather. Some wetlands dry out in summer, then fill with rain in autumn or when snow melts in spring. Summer dry periods are important in several ways. During that dry window, seasonal grasses and flowering plants can germinate on exposed mudflats; later in the year, they will provide important food and cover for wildlife. If a wetland dries up periodically, it will not support fish, which prey heavily on larval amphibians, including salamander and frog tadpoles; such seasonal or temporary wetlands are key breeding habitats for many amphibians. As years pass, some wetlands silt in and gradually become damp or dry habitats that lack standing water.

Beaver ponds often include wetlands, both because beavers seek out and use such areas and also because these large and industrious rodents often create wetlands when they build their stick-and-mud dams across streams. Beavers cut down and eat trees, mostly ones that are less than three inches in diameter. They generally build a series of ponds until they use up the available food in a stream drainage (it may take 10 or more years), and then they move on. Their dam will last for several more years as silt and organic matter build up behind it. Finally, during a spring thaw or a flood, the dam will give way. Most of the pond's water will drain out, leaving an open area where grasses and sedges will grow in the rich soil. Later, shrubs may seed in. Deer, bear, grouse, turkeys, flycatchers, and many other birds will use the beaver meadow, no longer a wetland but now a sunny, productive opening in the woods.

You can do several things to improve the habitat value of an active beaver pond and any associated wetlands. Put up wood duck nesting boxes on posts set in the mucky soil of the pond, with a metal predator guard ringing the post to keep raccoons and other predators from climbing up and raiding the nest. Erect bat roosting houses 12 to 20 feet above the ground in open areas near the edges of the pond. Bats will take shelter in the boxes during daylight and chilly weather. In some regions—including parts of the South where duck hunting is popular—landowners and hunting clubs manage beaver ponds to attract and hold waterfowl. The key is to control the pond's water level at different times of the year by adding a beaver-proof flow-control device that goes through the dam below water level. A popular one is the Clemson Beaver Pond Leveler. Drawing down the water to a depth of one to two feet in summer stimulates the growth of food plants around the pond's margins, including millet, duckweed, frogbit, smartweed, beak rush, bur-reed, and pondweed.

IMPROVING WETLANDS

Most approaches to improving wetlands are grounded in common sense and the principle that increasing habitat diversity will attract and support more and a greater range of wildlife.

Does a stream feed a wetland on your land? You can plant a buffer of native vegetation—grasses, wildflowers, shrubs, and trees—on both streambanks, 25 to 50 or as many as 300 feet wide. Buffers intercept runoff from farm fields, lawns, and roads, trapping soil particles, chemicals, fertilizers, and animal wastes and preventing them from flowing into the stream and then the wetland. Shade cast by buffer plantings helps keep water cool, limiting the growth of filamentous algae that can deplete oxygen. Well-oxygenated water will encourage insects and other invertebrate life, important prey for amphibians, reptiles, birds, and mammals. Buffers also function as wildlife travel corridors and breeding areas. Black ducks, mallards, gadwalls, and teal site their nests on the ground in thick grasses or low shrubs, and, as soon as their ducklings hatch, lead them to water. In lieu of planting a buffer, erecting a fence at least 25 feet from a stream or wetland will keep livestock from walking near and through the water, limiting water pollution and sedimentation and preventing bank erosion while abetting the growth of plants that wildlife use for food and cover.

You can also plant a buffer around a wetland. Shrubs and trees in a buffer can provide fruits and nuts that are not present in the wetland itself. Willows are fast-growing shrubs and trees that produce shade; deer and moose browse their twigs, and beaver, snowshoe hare, rabbits, and rodents feed on various parts of the plants. Birds such as flycatchers nest in the thick crowns of shrub willows. In buffer zones, keep or create snags as perches for birds and potential sites for cavities. Sycamores often develop cavities that wood ducks use for nesting. Build brush piles and leave hollow logs on the ground in buffers and along wetland edges. Planting clusters of conifers creates pockets of dense thermal cover that birds and mammals will use in winter.

Other ways of improving wetlands include thinning cattails when they become so thick that they don't allow wildlife to move about freely or other plants to grow. You can cut trees so their tops fall into wetlands; their branches will provide hiding cover for fish, and their trunks will become basking sites for turtles and water snakes. You can also plant areas of grass on drier ground nearby; alternating plots of cool-season and warm-season grasses, such as switchgrass, will add diversity. Putting in or exposing a strip of sandy soil on a warm, south-facing slope will give turtles a place to dig ground nests and lay eggs. Up to 80 percent of turtle nests are rifled and destroyed by raccoons, skunks, opossums, and foxes. If you suspect that turtles have dug nests in open soil near a wetland, you can protect them temporarily with wood-and-wire cages. Consult your state wildlife agency or university cooperative extension service, or search the internet for details on how and when to protect turtle nests.

RESTORING AND CREATING WETLANDS

You may be able to restore a wetland by removing, plugging, or disabling structures installed in the past to drain an area of ground and turn it into pasture or cropland. Many such acres are only marginally productive, and returning them to wetlands makes both ecological and economic sense.

To identify former wetlands, look for areas that have standing water in the spring. Contact your local USDA Natural Resources Conservation Service (NRCS) or U.S. Fish and Wildlife Service office. They will have maps that show whether your site has the hydric soils needed to successfully restore a wetland. They may also have maps showing whether and where the area was previously ditched or drained, including the location of underground drainage tiles. After visiting your site, biologists with those agencies may offer cost-sharing funds for projects, including voluntary fixed-term easements for conservation land.

Removing or disabling drains and ditches can restore the natural flow of water. Over time, the wetland may heal itself, usually becoming a shallow marsh. Emergent plants, such as grasses or sedges, may sprout from seeds that remain on the site. Wetland shrubs and trees would take much longer to return on their own, so planting native types from local sources can speed up the process of restoring a productive habitat.

When consulting with and learning from conservationists and habitat biologists, clearly communicate your goals, the amount of money you're able to spend, and what kinds of wildlife you hope to attract. They will guide you through the planning process, which usually includes surveying the site, designing the wetland (including its size, shape, and water depth), securing any needed permits, laying out the project on the ground, moving earth or disrupting drainage systems, and preparing the substrate and soil for the return of water and the growth of plants. They can help monitor wetland plants and offer advice on how to suppress any invasives that may show up. Both the Fish and Wildlife Service and NRCS work with private contractors who offer services such as laying out wetlands, grading, building impoundment structures, and planting native vegetation.

Here are some factors to consider:

Water Availability: The amount of water on your site, and the time of year it is present, will help determine what kind of wetland you end up with and the types of wildlife that will use it. On some wetlands, special devices can be used to regulate water levels. They can be simple, such as a spillway with removable flashboards, or a more complex setup relying on pumps.

Size: The larger a wetland, the greater the potential to support more diverse plants and wildlife. You'll probably end up with good diversity in a wetland of half an acre to five acres. Wetlands complexes that take in dozens or hundreds of acres are incredible habitats and fascinating wildlife hotspots.

Shape: A round wetland with a straight inflow channel will have a limited amount of shoreline. You'll attract more wildlife if you build or restore a wetland with an irregular shoreline. Features like coves, peninsulas, and islands add diversity, as do different slopes to the banks and various water depths. All of these features need to be planned before any construction begins.

Slope: A gentle slope to a wetland's banks means that more land will be exposed during low water periods, creating mudflats that support emergent plants such as grasses and rushes. Many birds (bitterns, for example) hide their nests among emergent plants. Shorebirds feed on insects and other invertebrates on mudflats. For most wetlands, a good slope ratio is 1:10, with the land losing one foot of elevation over every 10 feet of distance. Gentle slopes let amphibians and reptiles enter and leave a wetland more easily. To increase diversity, a wetland can be designed to have different slopes in different areas (although most of them should be gentle).

Water Depth: The water depth in most restored wetlands averages around 18 inches, which favors emergent plants. Herons, bitterns, and egrets hunt for prey in shallow areas. Dabbling ducks (also called puddle ducks), such as teal, mallards, and black ducks, float on the water and "dabble" with their bills to pick up floating plant foods or tip up their rear ends and reach underwater to nab food in shallows. Water depths from 18 to 48 inches support submerged and floating plants; fish can lay eggs in waters of those depths. In water four to six feet deep, fish can overwinter under ice, and reptiles and amphibians can burrow into bottom sediments and hibernate, absorbing oxygen through their skin. An irregular or undulating bottom in a restored or constructed wetlands will lead to water of differing depths.

Water Quality: A wetland will always be a better habitat if it is fed by good-quality water. Sudden changes in pH can harm wetland plants, and too much nitrogen and phosphorus can spur the growth of algae and lower oxygen levels in the water. Underwater plants need clear water so that sunlight can reach their leaves, letting them photosynthesize; turbidity from suspended sediments can make for cloudy water and unhealthy plants. A buffer zone around a wetland will help filter out excess nutrients and sediments, reducing turbidity.

Plant Structure: Wildlife usually select feeding, breeding, nesting, and resting habitats based on the density and height of vegetation rather than on specific types of plants. A wetland with an irregular shape and different water depths will encourage aquatic plants with a variety of structure and forms.

Dominant Plants: The amount and depth of water helps determine the species of plants that will do best. You can plant certain types of vegetation to favor specific types or groups of wildlife. If you rely on local vegetation growing back, including wetland plants whose seeds may remain on site, then the dominant plants in your wetland will likely be the same types as those in nearby wetlands. Study those plants and determine whether they'll attract the kinds of wildlife you want to encourage. You can also affect the type and location of plants in your wetland by changing its water level, for which you'll need some sort of a drawdown mechanism.

Interspersion: Interspersion is a technical term for the way in which areas of open water and different plant types mix in a wetland. Factors such as water depth and fluctuation, bank slope, and bottom substrate affect interspersion. A site where different kinds of plants grow to different heights, broken up by or woven through with areas of open water, will be more diverse than a wetland with only one type of plant or only open water. It will also be more beautiful, displaying a range of textures and colors. A mix of open water, mudflats, and zones of emergent plants and shrubs will attract a good mix of wildlife. Most waterfowl prefer wetlands that are about one-half open water and one-half emergent plants.

CREATING A VERNAL POOL

You can make a wetland on a site where no wetland existed in the past. So-called "constructed wetlands" are routinely built to treat wastewater, stormwater, acid mine drainage, and agricultural runoff. Landowners can also make wetlands with the goal of attracting and helping wildlife.

One type of wetland that most landowners can successfully create is a vernal pool. Up to half of all frogs and a third of all salamander species rely on vernal pools to reproduce. Mallards and wood ducks rest on their waters and feed on insects, crustaceans, and seeds during their northward migration in spring. Box turtles and garter snakes use vernal pools as feeding hotspots as they move about within their territories. Bats drink from vernal pools and catch flying insects in their vicinity.

Vernal pools are *ephemeral wetlands*, so-called because they dry out every year or in drought years. For that reason, they don't support fish, nor do they have many eastern newts—both of which eat amphibian larvae. Without those predators, tadpoles of wood frogs, spring peepers, spotted salamanders, and spadefoot toads can successfully metamorphose and emerge from vernal pools.

As with restoring a wetland, a good early step is to contact your local office of the U.S. Fish and Wildlife Service. Through their Partners for Fish and Wildlife program, this federal agency has helped tens of thousands of private landowners restore and create hundreds of thousands of acres of wetlands, including many vernal pools. Before you talk to the folks at Fish and Wildlife, you may want to learn more about vernal pools and how to create them through the writings of Thomas Biebighauser. A wildlife biologist and wetland ecologist in Morehead, Kentucky, Biebighauser has designed more than 5,000 wetland projects and built more than 1,800 wetlands in 22 states. His *A Guide to Creating Vernal Ponds* is available as a free download through the website of his company, Wetlands Restoration and Training. Biebighauser's *Wetland Restoration and Construction—A Technical Guide* is a more thorough 188-page textbook that can be ordered from Forestry Suppliers or Amazon.

Biebighauser points out that you don't need to be an engineer or a biologist to create a vernal pool. You do, however, need to understand how to build a pond that will hold water long enough so that aquatic plants can become established and for aquatic larvae, both insects and amphibians, to develop before the pool dries out. Water-permeable soils, a poorly constructed core under a dam built to hold water, and failing to compact the soil properly during construction are all reasons that vernal pools fail. Biebighauser stresses the importance of not building on a site that's already a functioning wetlands. He writes: "Some clues that may alert you to the presence of a seasonally dry wetland include: dark stained leaves, caddis fly larvae cases, fingernail clams, aquatic snails, bright-green sedges, and a lack of trees growing in the depression. These natural wetlands are most likely already providing habitat to many plants and animals"—yet another excellent reason to have a Fish and Wildlife Service wildlife biologist visit your property and help you find just the right spot for a vernal pool.

Fine-textured soils, such as clay and silt-loam, do a good job of holding water, while soils with too much gravel or sand do not. Biebighauser presents a simple technique to judge whether the soil on your site is impermeable enough to hold water: If you can mold or squish the soil between your thumb and forefinger and form it into a ribbon at least two inches long, it will likely hold water.

It's best to build a vernal pool in late summer or early fall when soils are fairly dry. If your site has silt loam or clay soil that extends down to rock or an impermeable soil layer, a bulldozer can be used to make a pool. For a level site, the bulldozer excavates a gentle, shallow depression and then will compact the soil heavily. If the site has some slope, you'll need to make a dam to trap water runoff. First, the dozer removes the topsoil and sets it aside. Then it constructs a core under the area where the dam will go—think of the core as an impermeable wall below the ground with the dam sitting on top of it. The core will prevent any water from leaking out beneath the dam. To build the core, the dozer digs a trench as wide as its blade. Then it pushes the silt loam or clay soil back into the trench and compacts it by moving back and forth repeatedly on top of each layer of fill. The dam is then built on top of the core, also by packing down successive layers of fill. (It often takes longer to build a proper core than to construct the dam on top of it, Biebighauser notes. Not understanding the purpose of a core, or failing to build one properly, results in a dam that won't hold water—another good reason to work through the U.S. Fish and Wildlife Service, who should be able to line you up with a dozer operator who has experience building vernal pools.)

The dam is graded to have a gradual slope, and the topsoil is spread back over the area that will fill with water and become the pool. To prevent erosion and make the site look better, you can seed bare soil using wheat or rye mulched with straw. Wheat and rye are annuals that will live less than a year, by which time natural plants should have begun to colonize the area. Biebighauser cites research done on the Daniel Boone National Forest in Kentucky in which scientists found more than 50 species of aquatic plants grew naturally, without being planted, within five years after vernal pools were built. Keep in mind that you will attract a greater variety of plants—and, ultimately, wildlife—if you site your pool near other wetlands.

You can also build a vernal pool on water-permeable gravel or sandy soils. The bulldozer excavates the area, and then a contractor installs a heavy-duty commercial-grade plastic liner of the type used beneath landfills. The liner can be protected on both bottom and top by a geotextile fabric pad to guard against puncturing, and then covered by six inches of soil.

For a small pool, especially if you can enlist family members and friends, and perhaps students in a local science class, you may be able to excavate a pond using shovels—after the digging is done, remove sharp sticks and rocks, lay down a heavy-duty liner, put the topsoil back, then seed and mulch. The next spring you will have a vernal pool, and in only a few years, you'll probably be listening to the calls of frogs and seeing dragonflies zip back and forth catching insects. Well-made pools with good liners have lasted 30 years and longer.

8

Special Habitat Features

THE SMALLER A PARCEL OF LAND, the less likely it will have abundant natural diversity to attract a range of wildlife. (For ways that you can increase diversity on a small property, see Chapter 10, "Habitat Around Your House.") However, if you own several—or several hundred—acres, significant diversity may exist on your tract. It's worth learning about various habitat features so that you can protect, enhance, or create them.

Wide-ranging mammals such as deer, coyotes, and black bears take advantage of *horizontal diversity:* different habitats and features spread out across the landscape, offering a variety of food and cover types. It's the same for larger birds, including hawks, falcons, vultures, crows, and ravens, all of which can fly long distances to find the food and cover they need. For forest songbirds, it's *vertical diversity* that counts. Most, and perhaps all, forest songbirds use more than one vertical layer in a woodland for feeding, roosting, breeding, nesting, and rearing young. In an article titled "What Do Animals Need, and Do

Rotting logs and branches provide physical structure at ground level, which can be refuges for invertebrates, amphibians, and small mammals. Ground-nesting birds often tuck their nests in under the sides of logs.

A small bear could hibernate inside a hollow tree as large as this sugar maple. Bats, raccoons, opossums, porcupines, squirrels, and owls are a few of the many animals that shelter in hollow trees and sizable cavities.

Your Woods Provide It?" Stephen Long, a former editor of *Northern Woodlands* magazine, writes: "Complex three-dimensionality is at the heart of habitat. And the more complex the structure of the forest, the greater [the] diversity of animals whose needs will be filled there."

Look around on your acreage for the following habitat features. If you decide to develop a map of your property, you may want to include them.

SNAGS AND CAVITIES

Snags are standing trees that are dead or dying. (A broken-off snag, especially one that's shorter than 20 feet tall, is called a *stub*.) Snags and their branches provide perch sites for birds, where they can see all around while carrying out important life tasks. Raptors such as kestrels and broad-winged hawks use prominent branches for *hunting perches*, as do kingbirds and other flycatchers. Kingfishers and bald eagles use snags near water as *fishing perches*. Songbirds use snags as *singing perches*.

Termites and the larvae of certain beetles feed on the wood of snags and snubs. Carpenter ants do not eat wood, but they tunnel into decaying wood to build their nests. Woodpeckers and nuthatches use their sturdy bills to dig out these insects, an important food, especially during winter. As snags deteriorate, their bark loosens; before it falls off, it may form a "bark cavity." Forest-dwelling bats such as the little brown bat, northern long-eared bat, and Indiana bat rest in such hideouts during the day. Brown creepers are small, inconspicuous forest birds that often build their nests in the space between loose bark and the trunk or inner wood of a snag.

Snags may develop hollow areas, called *cavities*, inside their trunks and branches. Many birds and mammals use cavities for nesting, rearing young, resting during cold or stormy weather, storing food, hiding from predators, or hibernating. Birds such as woodpeckers, nuthatches, and chickadees are *primary cavity users* that excavate their own cavities in wood. *Secondary cavity users* don't do their own excavating (their bills are specialized for uses other than chiseling wood) but seek out existing cavities in which to nest and lay eggs. They include wood duck, common goldeneye, saw-whet owl, screech owl, barred owl, kestrel, chimney swift, bluebird, wren, great crested flycatcher, and many more. Mammals that are secondary cavity users are bats, mice, marten, fisher, raccoon, porcupine, squirrels, weasels, and black bears. Reptiles and amphibians also shelter in cavities. Different kinds of wildlife use cavities of different sizes. Wood ducks and barred owls require large cavities for nesting. Tufted titmice and red-breasted nuthatches use smaller cavities.

Black bears use especially roomy cavities and hollows for hibernating.

If cavities don't exist on your land, you can build and put up nest boxes of various sizes and configurations that wildlife will use in much the same ways that they use natural cavities. (State wildlife agencies and wildlife organizations offer plans for a range of nest box designs.) You can also start a natural cavity yourself: Find a snag or an ailing tree whose heartwood has begun to rot but whose bark remains intact. Drill a hole about two inches in diameter through the bark and into the heartwood, placing it a few inches below a stout limb. Chickadees and nuthatches may spot the hole and use it as an entry point to begin excavating the rotting wood beneath.

Resist the temptation to cut snags for firewood. Leave these natural supermarkets and apartment houses for wildlife. If you have few or no snags on your land, you can create them. Select a live standing tree—preferably one with plenty of horizontal branches—and girdle it. Using downward blows with a sharp axe, cut a line about three-quarters of an inch deep all the way around the tree's trunk. Then hack a second line three or more inches above the first. Using the axe blade or a large chisel, cut and pry out the bark and cambium layer between the two lines. Within a few years, the tree will die—and it will enter a "second life" as a snag.

DEAD WOOD

Sooner or later, a snag will fall on the ground. Trees also get toppled by wind and ice storms, and branches periodically break and fall off. These join the stumps of cut trees to form the deadwood component on a forest's floor. In this moist ground-level setting, fungi and insects enter the wood and begin breaking it down. Centipedes, salamanders, frogs, toads, snakes, mice, and shrews find food and cover in crevices, cracks, cavities, and zones of rotten wood. Birds, skunks, raccoons, and bears feed on those smaller creatures.

Dead wood also provides physical structure in a forest or woods-edge setting. Male ruffed grouse hop up onto fallen trees and use them as "drumming logs": The birds grab hold of the wood with their feet and then beat their wings rapidly, making a repetitive low-frequency sound that carries over long distances and attracts females for mating. Grouse, turkeys, towhees, and other birds build their nests on the ground, often tucked in under the curve of a log. The top surfaces of fallen trees make convenient travel corridors for shrews, mice, chipmunks, and squirrels; timber rattlesnakes coil up next to logs, waiting for prey to scamper past.

You can cut down trees to create dead wood on the forest floor. If you conduct a timber harvest—especially an even-aged one, also known as a clearcut—leave some logs lying on the ground for wildlife. Owners of smaller properties, including house sites in woods, often make the mistake of "cleaning up" an area by removing or burning all of the "unsightly" fallen branches and rotting logs. When it comes to nature, neatness is the enemy of diversity: You will see a lot more wildlife if you let those features remain to provide food and cover as they slowly decompose.

STILT-ROOTED TREES

Yellow birch trees grow on cool, moist soils from the mountains of western North Carolina north to Maine and Minnesota. In winter, they drop thousands of small, lightweight seeds. Should a yellow birch seed land on top of a dead stump or log, the resulting birch seedling will send down roots that wrap around the dead wood while making their way to the ground. Years later, the stump or log will rot away, leaving a mangrove-like "stilt-rooted" tree. A "perched birch" offers a cavity space between its thick supporting roots—a great spot for an Appalachian or New England cottontail to wait out a storm, or for a salamander to find a cool, moist area in which to rest.

WILDLIFE SKETCH

A Perilous and Uncertain World

Medium-sized weasels, minks spend most of their lives in and near water. They also venture onto dry land, especially in winter when streams, ponds, and wetlands freeze over. Adults may hunt along more than a mile of stream corridor.
© *iStock.com/Dantesattic*

AT FIRST, I THOUGHT IT WAS A WEASEL. The slender predator bounded along the bank of Swift Run 30 feet away. Our attention immediately leaped from the towering hemlocks that filled the mountain valley to the lithe, dark creature as it snaked its long body between downed logs and over and around mossy rocks.

It was a mink—its body a uniform chocolate brown, its coat glistening in the diffused light that reached ground level. The mink was less than a foot long from its nose to the tip of its tail—probably a young animal, born in the spring and possibly on its own for the first time in its life. A life that might be brief, should a hawk or fox be watching, because the mink seemed oblivious to any danger as it ran unconcernedly and, I now saw, somewhat clumsily along the stream's edge.

My 12-year-old son, Will, had already scrambled his own binoculars out of his daypack. He and I followed quietly, keeping up with the mink. I put the back of my hand against my lips and made a squeaky kissing sound—in the past, I've called in weasels by imitating the high-pitched distress cry of a small animal like a vole or mouse.

The mink stopped and stood on its hind legs, facing in our direction. Then it hurried toward us through the ferns and moss and low plants and tree seedlings and woody debris. The youngster popped its head up 10 feet away and stared at us with bright black eyes. We must have looked at one another for 15 seconds. Then the mink turned, jumped over a fallen branch, dived into the creek, and swam to the far side—where it was joined by a second mink.

I had formed the opinion that the young mink was dispersing, leaving the area in which it had been born and raised. The second mink was larger, a paler shade of brown, obviously the youngster's mother. She ran straight to the pup. We heard faint, high-pitched chipping sounds. The mother licked the young mink on its neck and head. The youngster broke away and resumed loping down the hollow. The mother set off hunting, sticking her nose into nooks and crannies, checking between rocks. She darted beneath a green mat of sphagnum moss, struck with her muzzle once, twice, and emerged with a dark brown creature struggling in her jaws. She was too fast—even with the binoculars, I couldn't tell if she'd caught a rodent, a shrew, a salamander, or a frog. With the prey in her mouth, she sprinted off, following her pup.

Obeying a natural imperative to leave its home, the youngster was setting out to find a territory of its own. The mother mink was equally driven to continue nurturing and protecting her own flesh and blood.

It was one of those amazing wildlife encounters that you sometimes luck into by putting yourself in a rich habitat—in this case, a stand of old-growth hemlocks and pines protected within Snyder Middleswarth Natural Area in central Pennsylvania.

As I lowered my binoculars, I heard Will whisper "Awesome!"—one of the few times my son, the indoor dweller, the computer devotee, dragged along from one natural experience to the next from the time he was an infant, openly acknowledged the wonder of what he saw.

What went through my mind just then was the bittersweet understanding that my own son would someday leave his parents and his home and make his way into a perilous and uncertain world.

SPRING SEEPS

Spring seeps exist where water wells up from the ground to form damp depressions, trickling streams, or small pools. (Biologists call these habitats "discharge wetlands.") Seeps often occur on hillsides or at the bases of mountains. Grasses, sedges, ferns, mosses, and flowering plants grow in seeps. Insects and other invertebrates living in seeps offer high-protein food to wildlife. The groundwater that creates seeps usually stays around 50 to 60 degrees Fahrenheit year-round. Spring seeps are especially important in winter, when they may be the only accessible source of food and drinking water in an area.

Deer, elk, and moose sometimes visit seeps for elements such as calcium or sodium that may be present in the groundwater and the saturated soil. Frogs and salamanders live in seeps; they attract predators such as skunks, raccoons, and mink. Ruffed grouse eat leaves of low plants growing in seeps. When powdery snow lies deep, seeps may be the only places where wild turkeys can walk and feed. When American woodcock and robins migrate north in early spring, they find worms and other invertebrates in seeps when cold spells freeze the ground or snow covers other foods.

The best way to manage a seep is to protect it from anything that might harm it or its plant and animal life, such as herbicides or insecticides. You can cut back competing vegetation around seeps to favor food-producing plants and shrubs, and plant moist-soil native shrubs along sunny fringes. When logging, keep a buffer of trees around seeps so sediment doesn't wash into them. Keep treetops and logging debris away from paths that wildlife use to visit these small but valuable habitat features.

SEASONAL POOLS

Seasonal or temporary pools are small wetlands that fill with water following autumn rains or spring snowmelt. *Vernal pools* appear in spring; *autumnal pools* fill up in fall and remain filled with water in spring. Spring is the best time to look for seasonal pools—and to listen for them when the calling of spring peepers, wood frogs, and other amphibians proclaims their presence. Vernal pools are magnets for these and many other amphibians, which come crawling or hopping to these breeding hotspots from the surrounding terrain. Seasonal pools can support a fascinating web of life that includes aquatic insects, fairy shrimp, and clam shrimp. Larger pools attract turtles, snakes, waterfowl, wading birds, songbirds, and large and small mammals.

If you have a seasonal pool on your property, take good care of it. When logging, leave a buffer zone around this high-value habitat feature, and make sure no agricultural runoff gets into it. Chapter 7, "Wetlands for Wildlife," explains more about vernal pools, including how to build them.

RIPARIAN ZONES

"Riparian" comes from the Latin word *ripa*, for "riverbank." Riparian habitats exist on the banks of rivers and streams. They often have rich soil laid down by past flooding, promoting lush plant growth that offers wildlife dependable food, cover, and travel corridors. More than 80 kinds of birds feed and nest in streamside vegetation. Turtles and snakes frequent riparian zones, as do mammals such as otters, fishers, mink, muskrat, beaver, foxes, rabbits, and many more.

A *riparian buffer* is a strip of vegetation that helps shade a waterway and protects it from surface runoff that may contain sediments, excess nutrients, or pollutants. Riparian buffers also limit erosion and help stabilize stream banks. In farming areas, conservationists recommend creating a buffer at least 50 feet wide—and made up of three zones—on each side of a stream.

The first zone, next to the stream, should have native trees such as cottonwood, black willow, hackberry, magnolia, honey locust, black gum, sycamore, silver maple, black walnut, bur oak, and swamp white oak. Contact your state wildlife agency, county extension service, soil conservation district, or the USDA Natural Resources Conservation Service to learn which trees grow

Rich soils along rivers and streams promote lush plant growth, which translates into food and cover for wildlife. In farming areas, conservationists recommend creating buffer zones at least 50 feet wide on both sides of streams.
Tom Berriman

well in bottomlands and riparian zones in your state or region.

The middle zone of a buffer should be native shrubs and smaller trees. Haws, viburnums, crabapple, elderberry, hawthorn, American plum, redbud, hornbeam, serviceberry, dogwood, and winterberry do well in such settings.

The outermost zone should be mostly native warm-season grasses, which slow runoff and absorb contaminants before they reach the two inner zones.

Once planted and established, riparian buffers usually need little maintenance. Through its Conservation Reserve Enhancement Program, NRCS offers funding to landowners, including farmers, to plant riparian buffers, paying an annual rental fee for every acre taken out of agricultural production and put into a buffer-zone habitat. Another worthwhile practice is to fence off streams from livestock. If necessary, leave stream crossings in a few areas so farm animals can access the stream for drinking or crossing it to get to another field.

On a smaller scale, plant clumps of food-producing shrubs at intervals along a stream on your land. If such shrubs are already growing, clear away taller, less-valuable vegetation to let sunlight reach the shrubs and boost their growth and productivity. Where non-native invasive shrubs choke riparian zones, suppress these undesirable plants by pulling them up, cutting them back repeatedly, or carefully targeting them with herbicide. (See Chapter 9, "A Plague of Invasives" for more detail.)

APPLE TREES

Many folks don't realize that apple trees are non-native aliens—although not *invasive* aliens, like so many exotic plants that are elbowing their way into, and too often dominating, natural habitats. Apple trees came to the New World with the first European settlers. Today, we still relish eating apples—and so do wild animals, which seek out the fruit on and beneath apple trees growing in untended orchards, fencerows, thickets, and old fields.

The familiar red, yellow, or green fruits ripen in late summer and fall. Deer, bear, coyote, foxes, raccoon, porcupine, opossum, squirrels, chipmunk, woodchuck, crow, woodpeckers, pheasant, ruffed grouse, blue jay, robin, thrushes—all avidly eat apples in years when the trees produce fruit. Deer browse twigs and leaves. American woodcock feed on worms in the rich soil that builds up beneath apple trees. Sapsuckers drill rows of "feeding wells" in the trees' bark, then eat insects and other invertebrates drawn to (and sometimes trapped in) the sap that wells up to fill the small pits. In winter, ruffed grouse nip off energy-packed buds in the trees' crowns, and rabbits gnaw on the bark. In spring and summer, many songbirds nest in the dense crowns of apple trees; other birds use cavities in the trunks or limbs of older specimens.

To "wake up" an old apple tree, cut down nearby taller trees casting shade on it, especially any that stand to the south of the apple tree. This procedure, known as releasing or "daylighting" the tree, will cause it to produce more fruit. Prune a wild apple tree by removing any dead wood. Cutting out about a third of the tree's live wood will cause it to pull more nutrients from the soil, put out new growth, and produce more apples. (To minimize injury to the tree, prune any live wood in late winter.) Spreading manure and compost under a tree will also boost its health and productivity. A more drastic approach is to chainsaw off the top third of the tree's branches. Trees that have grown from local seeds are notoriously tough and will likely benefit from such rough treatment. Your local NRCS office or state wildlife agency may pay you to release and reinvigorate apple trees on your land.

FENCEROWS AND TRAVEL CORRIDORS

Fencerows get their start when someone puts up posts and strings wire between them; birds then

Fencerows help low-mobility wildlife live in heavily farmed areas by letting animals shift between one habitat patch and another, and from one local sub-population to another. Fencerow trees and shrubs also provide food.

perch on the fence and deposit seeds in their droppings. Over time, the area under and around the fence becomes overgrown with grasses, weeds, wildflowers, vines, shrubs, and small trees. Species such as hawthorn, wild plum, elderberry, sumac, blackberry, dogwood, fire cherry, pasture rose, wild grape, greenbrier, and red cedar provide valuable food and cover. (All too often, aggressive non-native types such as privet, honeysuckle, and multiflora rose show up as well.) Large trees, including nut-bearing oaks and hickories, or fruit trees such as black cherry and black gum, can be important components in fencerows.

Fencerows offer hiding cover to woodchucks, squirrels, rabbits, weasels, quail, pheasants, lizards, snakes, and other animals, which move within them or sneak along next to them, ready to duck into the fencerow's cover should danger threaten. Fencerows help low-mobility wildlife persist in agricultural areas by letting individual animals shift between one habitat patch and another, and from one local wildlife sub-population to another. Woodchucks often dig burrows in fencerows—holes that are also used by many other kinds of wildlife, including smaller rodents and snakes. Foxes and coyotes take over woodchuck burrows and enlarge them for use as dens. Open-country birds such as hawks, kestrels, shrikes, and kingbirds perch in fencerow shrubs and trees and sally forth to catch insects and small mammals above and in nearby fields. Unfortunately, fencerows have become an endangered feature, as farmers have removed many of them to increase field size so that bigger machines can work the land.

If you have a fencerow on your property, preserve it and consider expanding it. You can create a similar corridor by creating a strip of food- and cover-producing plants across a pasture or other open or mowed area. Such a corridor can link two separate habitat areas, such as a pair of woodlots or a woodlot and a shrubland. Biologists recommend that such corridors be at least 50 feet wide and up to 200 feet wide if you have the space. For a 50-foot corridor, plant a 20-foot-wide center strip of larger trees; nut-bearing types such as oaks and hickories (available from both state and private nurseries) work well, although they are slow-growing. On both sides of the tree strip, plant a shrub strip five feet wide to provide food and cover. Then to the outside of the shrub strips, plant 10-foot-wide strips of warm-season grasses and wildflowers.

Every 100 to 300 yards along the corridor, build a brush pile—or, if you have access to heavy equipment, push together some large rocks, leaving nooks and crannies between them. You can also establish "islands" of dense native shrubs and small trees—patches 30x50 feet up to 100x100 feet, perhaps focused on an old rock pile. These islands will help prey animals escape more easily from predators and make the fencerow a more effective corridor.

Beavers dam streams, backing up water that can cover one to 100 acres. Waterfowl use beaver ponds, and birds such as kingfishers and swallows find food in and above the water. Beaver ponds and beavers' feeding activities kill trees, creating openings in forested areas.

If non-native shrubs have infiltrated your fencerow, gradually reduce their numbers by cutting them back repeatedly, uprooting them, or treating them with herbicide. Plant and otherwise encourage native shrubs that offer better food for wildlife. If your fencerow is completely given over to invasives, don't just wipe it out and start over: It's better for wildlife if you leave the overall structure of the fencerow intact as you patiently work to swing its composition from non-native to native shrubs.

BEAVER PONDS

Our largest rodents, beavers are common and often abundant in rural and forested areas throughout the East. Using sticks, mud, and leaves, beavers build dams across slow-moving streams, backing up water to create a pond that can flood one to 100 acres, with a stick-built structure, called a lodge, that houses a mated pair of adults and their offspring. Beavers also live along rivers, on timbered marshy land, and on forest-edged lakes. They use their strong incisors to cut down trees of many species, then eat their inner bark, buds, and tender twigs and use the remaining wood to build and maintain their dams.

After a dam is raised, a habitat area will change from a woodland to an open pond, banishing some forms of wildlife while providing excellent food and cover for other species. If beavers move onto your land, you may consider yourself fortunate—or, perhaps, badly inconvenienced, should their dam flood a road, a farm field, valuable timber, or a Christmas tree plantation. (Learn more about beaver ponds in Chapter 7, "Wetlands for Wildlife.")

MAN-MADE PONDS

Many landowners have ponds of varying sizes on their properties. Some pond owners groom and mow areas of cool-season grasses—essentially lawns—right up to the banks of their ponds. This makes for a neat appearance, but if helping wildlife is your goal, a better approach is to plant and maintain a 15- to 20-foot-wide zone of native grasses and wildflowers all around the pond. Columbine, butterfly weed, aster, partridge pea, joe-pye weed, bee balm, black-eyed Susan, and goldenrod, plus the native grass little bluestem, combine to make a good buffer zone. Swamp milkweed, blazing star, and cardinal flower do well on damp sites. An unmowed buffer will discourage Canada geese (a nuisance species in many areas) from using the pond. Planting a few native shrubs and trees will add visual interest and diversify the food available to wildlife. Keep livestock and poultry out of pond buffers so they don't trample or eat vegetation, erode the pond's banks, and add their wastes to the water.

Buffer zones slow down runoff into the pond, trapping and filtering out leaves and grass clippings, soil particles, and elements such as nitrogen and phosphorus, all of which contribute to pond *eutrophication*, when overabundant nutrients spur plant life that then uses up oxygen in the water—oxygen needed by aerobic bacteria to break down waste materials and keep the pond ecosystem functioning with a healthy range of life forms: plankton, zooplankton, small insects, larger insects, fish, frogs, and waterfowl, on up to predators such as otters, raccoons, and ospreys.

Invasive alien plants that can take over ponds include purple loosestrife, flowering rush, phragmites, and Eurasian water milfoil. Herbicides and biological controls are an option for limiting some species. Cattails are native plants that can also choke ponds. Consult with your state's cooperative extension service for the best ways of preventing and dealing with invasive plants in ponds. Searching the internet for "best management practices for small ponds" will bring up downloadable guides from different states.

Thinking about building a pond? You can get good advice from your state's fish and wildlife agency, a nearby U.S. Fish and Wildlife Service office, or the USDA Natural Resources Conservation Service; the two federal agencies may also offer cost-sharing funds. When laying out a pond, keep in mind that an irregular shoreline offers more and better habitat than a straight or curved shoreline. If yours is a small property, you may still be able to put in a small pond to help wildlife. NRCS, the National Association of Conservation Districts, and the Wildlife Habitat Council promote a Backyard Conservation program. They explain small-scale conservation practices through a series of "tip sheets," one of which is titled "Backyard Pond."

FOREST OPENINGS

Openings and clearings break up the uniformity of a woodland and offer wildlife different foods than the surrounding forest. Openings can be created by natural disturbances such as fires, insect or disease outbreaks, floods, ice storms, and strong winds. Winds and lightning strikes cause small openings known as *forest gaps:* a single tree felled, which then knocks over several additional smaller trees. Low, light-loving plants will grow in these openings if deer don't suppress them with their browsing.

You can create openings and clearings by constructing wide logging roads and log landings (sites where logs are grouped for loading onto trucks during a timber harvest).

Most ecologists don't advocate creating forest openings in areas where woodlands are already fragmented—broken up with roads, trails, homes,

Landowners can plant grasses along woods roads and on landings where logs are grouped during timber harvests. Wild turkey hens bring their poults to grassy clearings to hunt for insects.
© iStock.com/WilliamSherman

farm fields, and utility corridors, and where plenty of edge habitat already exists. However, if your goal is to increase numbers of edge-dependent wildlife, especially game species such as quail, ruffed grouse, turkeys, and rabbits, you can manage 5 to 10 percent of your woods as openings. On a 100-acre wooded property, that means five to 10 acres of open land: five to 10 one-acre openings, or two to five two-acre openings.

The best spots to create openings are where there are already only a few trees growing, such as old fields that have not yet gone back to forest, places where the soil is poor or thin, well-drained droughty spots, rocky areas, and frost pockets where cold air suppresses tree growth. Openings should generally be twice as long as they are wide to provide close-by escape cover for animals that venture out into the cleared area. You can plant openings, such as log landings, with native grasses and flowering plants such as Canada wild-rye and little bluestem. In early spring, wildlife will head for openings on south-facing slopes, because edible plants will green up there first, soon after the snow melts.

Add fruit-bearing native shrubs to attract more and a greater variety of wildlife. An opening a half acre to two acres in size will let both sun-loving and shade-tolerant plants thrive. Plant sun-loving species, such as hawthorn, sumac, crabapple, and gray and silky dogwood, in the middle and on the north side of a clearing, since those areas get the most sunlight. Types that are more shade-tolerant—beaked hazelnut, flowering dogwood, serviceberry, and highbush cranberry—will do better in the partial shade cast by trees bordering the south side of the clearing. Openings smaller than a half acre are too small to get much sun and generally will support only shade-tolerant plants.

Most forest openings are short-lived features because the forest is always ready to fill them back up again with new, young trees. You can let an opening progress in its natural succession toward forest just as far as you want it to go. When the cover has reached the age or growth stage that you want, set it back by cutting or brush-hogging small trees and shoots.

MAST

"Mast" is an old English word for nuts that fall on the ground and provide food for wild and domestic animals. Ecologists classify nuts as *hard mast*, while fruits, berries, catkins (dangling flower spikes of trees such as aspens and willows), and buds are known as *soft mast*. Both types are important to wildlife.

Oak, hickory, black walnut, butternut, and beech trees produce hard mast of high quality. Black cherry is an important producer of soft mast. Other key mast-producing species are dogwoods, mountain ash, maples, persimmon, serviceberry, black gum, hawthorn, and American holly. Various reptiles, birds, and mammals feed directly on mast, eating it as appears on trees and shrubs, gleaning it after it has fallen to the ground, and sometimes storing it for future use. Mast from native trees and shrubs also provides food for insects (both adults and larvae), themselves a key source of protein for wildlife.

If mast trees are few on your property, consider planting some. Oaks are often a good choice, as there are a variety of species that do well in both dry and moist soils. Many privately owned forests have mast trees, but often, they're competing with same-age trees of other types that don't produce as much mast or mast of equally high value to wildlife. Through careful thinning, you can remove the less valuable trees that are hemming in or overtopping mast trees. Once freed from this competition, the mast trees will expand their leafy canopies and produce more food.

ROCK LEDGES, ROCK PILES, AND TALUS SLOPES

Rocky features add important structure to habitat. They offer cracks and crevices that provide cover for salamanders, snakes, bats, raccoons, rabbits, chipmunks, cliff swallows, phoebes, vultures, and many other kinds of wildlife. Ravens and peregrine falcons site their nests on cliffs; phoebes nest under rock overhangs. Rattlesnakes bask on rocks in sunny settings in spring; they also overwinter in deep spaces beneath jumbled rocks and areas of talus (piles or fields of stone that accumulate on slopes below cliffs or other rock sources). Rare flowering plants often grow on the soil around rocks of various geological compositions. Your state's wildlife or environmental conservation agency may have information about important plants that grow among different types of rocks in your area.

Making a rock pile in a shrubland, fencerow, or riparian zone adds an important structural element to a habitat. Plant food-producing vines or shrubs among the rocks.

9

A Plague of Invasives

BOTANISTS ESTIMATE THAT HUMANS HAVE INTRODUCED 50,000 PLANTS to North America over the last several centuries. Around 5,000 of these plants have gone wild and now compete—some of them very successfully—with our 17,000 native plants. Many exotic species cause few problems and just seem to "blend in": Dandelions, for instance, have been around so long they've become practically invisible (except when they flower in spring) and seem to be part of the local flora. Some, like wild apple trees, provide wildlife with nutritious food and good hiding cover. However, many others contribute to what has become one of the most challenging ecological problems in our land today: the takeover and degradation of natural habitats by invasive, non-native plant life. "Plague" is not too strong of a word to describe what these plants have brought—and continue to bring—to nature in the East.

Invasive plants thrive in America because they do not face diseases, insects, and browsing animals that keep them in check in their homelands. When songbirds eat the scarlet fruits of autumn olive, they spread the seeds of this exotic shrub far and wide.

A plant can be considered "invasive" if it spreads over a large area, severely suppressing or excluding other vegetation. Native plants can sometimes become invasive, such as cattails choking a pond, striped maples or hay-scented ferns taking over a recently cut forested tract, or wild grapevines smothering trees in a woodlot or timber stand (something you won't like if your goal is to grow big, commercially valuable trees, but probably *will* welcome if you want to provide food for ruffed grouse and songbirds).

Most of the plants we know as invasives are non-native species that succeed here because they do not face the diseases, insects, and browsing animals that keep them in check in their lands of origin. In eastern North America, some of the more common (and infamous) invasives include low-growing plants like Japanese stiltgrass and garlic mustard; vines such as Oriental bittersweet, mile-a-minute weed, and kudzu; shrubs such as multiflora rose, Japanese barberry, autumn olive, and bush honeysuckle; and trees such as smooth and glossy buckthorn, Norway maple, and ailanthus. Some invasives, including wild parsnip and giant hogweed, have sap that can severely burn people who pull them up without wearing protective gloves and then are exposed to sunlight before washing the plants' sap off their skin. Others, such as poison hemlock, can be fatal if ingested.

Some of today's invasives are ornamental plants that have escaped cultivation and outcompeted our native plants in a wide range of different habitats. State and federal conservation agencies must accept a large portion of the blame for our current situation: Back in the 1950s and '60s, they urged landowners to plant what were then viewed as "super plants" to help wildlife, including ones that have become hugely problematic, such as kudzu, multiflora rose, bush honeysuckle, and autumn olive. It's true that some invasives provide effective cover, thick vegetative structure that deer, wild turkeys, cottontail rabbits, ring-necked pheasants, and other animals use for resting, nesting, and hiding from predators. Some offer food of varying quality in the form of nectar, browse, or fruits. However, the native shrubs and other plants that those exotics have displaced—plants that our wildlife and insects evolved alongside—generally offer much better food and equally protective cover.

As well as having few natural enemies, the most successful invasives employ several biological strategies that help them take over habitats. Many produce copious fruits or seeds. The seeds of some invasives catch in the fur or feathers of small mammals and birds and are transported to new areas. Lightweight seeds and winged seeds such as those of Norway maple and ailanthus spread on the wind; others are carried by flowing water. People help them spread by accidentally picking them up on our machines and even our footwear and carrying them to new places. Birds eat the fruits of autumn olive, honeysuckle, white mulberry, and other invasives, then deposit the seeds in their droppings. You can often find invasives growing along fencerows and field edges and beneath trees in fields—all places where birds perch.

Some invasives spread vegetatively through their roots. Kudzu, Oriental bittersweet, common reed (also called phragmites), and Japanese knotweed have massive root systems that constantly send up new shoots, helping these plants quickly seize sites and produce monocultures in which a single plant species dominates an area and excludes all or most other plants. Branch tips of barberry and multiflora rose send down roots where they touch the ground. Another advantage held by invasives is that browsing animals such as deer and rabbits don't like how they taste and prefer to feed on native plants. Likewise, many insects do not feed on the leaves of invasives. Some invasives, including barberry, alter the soil's chemistry so that native plants can no longer grow. Because they draw nitrogen from the air

Oriental bittersweet outcompetes and smothers native plants that offer better food and equally thick cover. Since our insects do not eat bittersweet leaves, fewer insects will be produced in a habitat patch degraded by invasives, yielding less food for wildlife.
Lisa Wahle

and "fix" it in the soil, autumn olive and Russian olive can increase soil nitrogen levels beyond those tolerated by native plants.

When invasives take over a site, they cause serious harm to both native plants and animals. Oriental bittersweet can swarm over shrubs and climb 60 feet into trees, shading out foliage and adding weight that, combined with snow buildup, can break limbs or even bring a tree down. Dense stands of phragmites and purple loosestrife on the edges of ponds and wetlands crowd out small mammals and birds and outcompete native plants that provide important food for ducks and other waterfowl. Bush honeysuckle and buckthorn limit the growth of tree seedlings and reduce the abundance and variety of low plants.

During the autumn migration, songbirds burn huge amounts of energy as they wing southward to their wintering grounds, hundreds or even thousands of miles away. Native shrubs, especially dogwoods, produce high-quality fruits with ample fat that help songbirds replace calories they burn during flight. The fruits of invasives, often brightly colored, also attract songbirds, but they're low in nutrients: junk food for migrating birds. Diets based on invasives' fruits can cause songbirds to lose weight and have a poorer functioning immune system. Some invasives poison birds outright: Heavenly bamboo, an evergreen shrub popular in the Southeast for its bright autumn foliage and fruit, has cyanide-laced berries that have killed cedar waxwings.

Healing a Shrubland Helps an Endangered Rabbit

A tracked machine quickly uproots a big autumn olive on Cottontail Farm.

A SMALL YELLOW MACHINE CLANKED ALONG ON ITS STEEL TRACKS. It was a bright day in May 2012. The machine stopped in front of a sprawling autumn olive that was 10 feet tall and 15 feet across—one of the dozens of exotic invasive shrubs choking an old pasture on the aptly named Cottontail Farm near Scotland, Connecticut.

The operator set the machine's blade against the autumn olive and pushed. With a barely audible *crack*, the shrub tipped over. The operator opened a movable part of the blade, clamped onto the shrub, and shook the dirt from the autumn olive's roots. Then the yellow machine went clanking off with the uprooted olive, put it into a pile, and moved on to the next patch of unwanted shrubbery.

Tom McAvoy, the owner of Cottontail Farm, stood watching. Beside him was Ted Kendziora, a U.S. Fish and Wildlife Service biologist from Concord, New Hampshire. "We can make much better habitat for wildlife than what these invasive shrubs are providing," Kendziora said. He works with many landowners like McAvoy through the Fish and Wildlife Service's Partners for Fish and Wildlife program. Kendziora concentrates on developing projects in the six states within the geographic range of the New England cottontail, a rabbit whose population has been falling for decades as the shrubland and young forest habitats that it needs have been covered with houses, degraded by invasive shrubs, or eliminated by trees rising and spreading their leafy crowns and preventing sunlight from reaching ground level.

Kendziora and McAvoy walked over to where the autumn olive had been yanked out. The ground was almost bare—no little green plants for rabbits to munch on. They'd all been shaded out by the autumn olive. Kendziora pronounced the spot "an instant planting site. We'll smooth out the soil, scatter a bunch of native shrub seeds, and plant grass to prevent erosion while we wait for the shrubs to grow.

"The management plan for this site also calls for planting native shrubs—red mulberry, chokeberry, and dogwood. We'll put in 52 clusters of nine shrubs apiece in this 5-acre field. Add some fencing around the shrubs to temporarily exclude the deer, and this field will be well on its way to becoming a much better home for cottontails." Other shrubs that would be part of the restoration project included arrowwood, nannyberry, elderberry, spirea, and juniper.

McAvoy said that he looked forward to the improvements scheduled for his property, including creating native-shrub habitat on five fields and harvesting

At Cottontail Farm in Connecticut, Tom McAvoy fenced this planting of native shrubs to protect them from browsing deer. Trees in the background have been invaded by Oriental bittersweet vines.

eight acres of timber in a nearby 18-acre woodlot. The timber harvest would bring in income to balance some of the costs of shrub planting; it would also create conditions where young regrowing hardwood trees, plus pokeberry, blackberry, and blueberry shrubs, would create more dense cover for cottontails. McAvoy, a lifelong outdoorsman and hunter, noted that deer would use the new young-forest habitat as well.

McAvoy has hunted deer on the farm for years, including decades ago when a friend's grandfather owned the 115-acre property. When the land came on the market, McAvoy bought it. He and his wife moved into the 250-year-old farmhouse. Nowadays, McAvoy's sons, in their early 30s, join him every fall to hunt deer on the farm. "They're very conservation-minded," McAvoy said. "We've been hunting here together since the boys were young."

Photos from the 1960s show that the land was completely open. "This was an active dairy farm," McAvoy said. (He still keeps 30 acres in corn and soybeans and 10 acres in hay, which is used to feed the Highland cattle he raises.) However, as the years passed, dairy operations ceased, and the untended fields were invaded by autumn olive, bush honeysuckle, and multiflora rose.

That the aggressive shrubs afforded at least some cover to the local fauna was proven by the fact that biologists working for the state of Connecticut found and radio-collared New England cottontails on the McAvoy farm, as well as on several neighboring properties. The biologists study the rabbits' home ranges and habitat preferences, how far they move in different seasons, and how long individuals survive in the wild. Rabbits live-trapped on the farm have been sent to Roger Williams Park Zoo in Providence, Rhode Island, where a captive breeding program produces New England cottontails that are released to bolster existing populations elsewhere in the species' range and to start new populations in areas where biologists and landowners have created shrubland and young-forest habitat.

Kendziora calls himself a "hands-on habitat biologist." He's not reluctant to jump on an excavator and grub out invasives or pick up a shovel and plant native shrubs. "I try to work closely with landowners during the planning stage for habitat projects," he said, "and then later with the professional contractors who do most of the actual habitat-improvement work.

"Here on Cottontail Farm, I plan to work with Tom to restore the habitat gradually over the next few years. We want to make sure the existing habitat isn't disrupted too badly. We'll keep plenty of shrubs in place—including invasives—while we work to swing the balance from non-native species to native species that offer much better food and cover to rabbits and other wildlife."

A return visit to the farm in 2017 showed that considerable progress had been made. Monarch butterflies were feeding on the flowers of milkweeds and goldenrod that had sprung up naturally in the rehabilitated fields. When McAvoy first saw all the weeds coming up, he "called Ted up and said, 'Hey, I think we've got a problem.'" He laughed. "Ted replied, 'No, that's just what we want!'"

McAvoy began seeing "good numbers of deer and turkeys," along with many songbirds, hawks, and other birds. "It seems there are rabbits everywhere," McAvoy said. "Not long ago, I saw a big bobcat walk across the road heading toward the area where we harvested timber to make that stand of young forest.

"Predators like bobcats are part of the system, too. The rabbits are an important food for predators. If the habitat is good, the cottontails can mostly avoid the predators," McAvoy said.

Over the last seven years, McAvoy has received more than $50,000 in cost-sharing grants from the USDA Natural Resources Conservation Service for seven separate habitat-improvement projects, for which he has paid about 75 percent of the overall cost. McAvoy, a banker and an estate planner by profession, pointed out that the cost-share funding is considered taxable income. "That's something a landowner needs to take into account," he said. "They also should realize that they incur a significant obligation when they sign up for cost-sharing funding. I mean, you sign a contract. There's a lot to learn, a lot to understand."

Although private contractors handled the bulk of the habitat-improvement work, McAvoy and his sons also pitched in, planting and watering shrubs, putting up deer fencing, and building loose piles of rocks grubbed out of the fields, with crevices that cottontails can use for hiding cover. "For some of the work, I hired kids from our church," McAvoy said. Throughout the process, he received advice from the Fish and Wildlife Service.

Remnant populations of New England cottontails exist in parts of Maine, New Hampshire, Massachusetts, Rhode Island, Connecticut, and New York east of the Hudson River. Habitat created by landowners like McAvoy, along with similar projects on public lands undertaken by federal and state agencies, led the U.S. Fish and Wildlife Service to conclude in 2016 that the New England cottontail does not need to be classified as "threatened" or "endangered" under the federal Endangered Species Act. McAvoy was an invited speaker at a ceremony in southern New Hampshire marking that achievement. McAvoy has even traveled to Maine and Colorado to speak about his habitat work to other landowners thinking about helping cottontails and other wildlife.

"I admit that at the beginning I didn't fully comprehend the commitment I was making," McAvoy said. "Along the way, I learned a great deal. In the end, it all turned out to be a very positive and encouraging experience.

"It just makes you feel good to be able to help the rabbits, especially when you know that so many other kinds of wildlife—like the songbirds and the bobcats and the butterflies—are using the same habitat. I look at this as a legacy project, one that my sons will be part of in the future."

He added, "I really like it in the mornings when my wife and I sit on the porch and watch the rabbits coming out of the thickets to feed on grasses in the pastures. They chase each other around. To me, it looks like they're having fun."

If you want to help wildlife on your land, familiarize yourself with invasive plants and commit to making a serious effort to control them. Unless you live in a pristine, deep-woods setting, you will probably have at least some invasives on your property. It's almost impossible to get rid of all invasives, but if you just turn your back and ignore them, there's a good chance they'll end up overrunning your natural area.

BATTLING INVASIVES

Douglas Tallamy is an entomologist and wildlife ecologist at the University of Delaware and the author of two books on how to help wildlife by planting native shrubs and other plants. In a forward to the second edition of *The Woods in Your Backyard* (a publication of several university extension services), Tallamy writes: "Studies of animal food webs have shown that the insects supported by native trees, especially various caterpillar species, directly or indirectly provide the protein and fats that sustain most amphibians, reptiles, mammals, and birds." He continues: "Encouraging native tree and shrub species is a challenge with the onslaught of invasive species that are displacing native plants," and suggests that landowners can improve local biodiversity by "supporting native plants, killing non-native plants, and controlling deer populations that give invasive plants a competitive advantage over the natives that deer love to eat.

"You can think of this as active land management, or simply gardening; in either case, we can no longer simply 'let nature take its course' and expect the return of productive ecosystems."

Before developing a habitat-improvement plan for any property—or taking measures that will change its vegetation, such as letting a field grow up in shrubs or even logging a woodland—learn to identify invasive plants likely to live in your area. There are many online resources to help you figure out whether those bushes in your fencerow are wineberry shrubs (an invasive from Asia) or native blackberries or raspberries, and whether that smallish tree in an old field is a black locust or an alien buckthorn. One good guide is *Mistaken Identity? Invasive Plants and their Native Look-Alikes*, published in 2008 by the Delaware Department of Agriculture and the USDA Natural Resources Conservation Service, which is available online.

In 2017, Penn State Extension brought out a compact spiral-bound handbook with photos and basic details of 25 of the most common invasive grasses, herbs, shrubs, trees, and vines in the East. *Invasive Forest Plants of the Mid-Atlantic* costs $10; preview it by searching for it on the internet, then download a PDF. The book includes "Treatment Calendars," showing which months represent the best times of the year to use various techniques for fighting invasives. For more detail, consult the book *Invasive Plants: A Guide to Identification and the Impacts and Control of Common North American Species*, by Sylvan Ramsey Kaufman and Wallace Kaufman (Stackpole, 2007).

The website of the National Invasive Species Information Center presents a wide range of information. Hosted by the U.S. Department of Agriculture, the site includes news about invasives; species profiles; and up-to-date maps generated by the Early Detection and Distribution Mapping System showing the presence of different invasives at the county, state, and national levels. The Nature Conservancy's website is another excellent resource. From TNC's main website, go to "North America" or "United States," then use the site's search engine to search on invasives in general, specific invasives, or related topics, such as "native plants."

Thoroughly and systematically search your land and record where invasives are growing. Early spring is a good time to look since many invasives put out leaves before native plants. If you find few invasives on your land but a neighboring tract has plenty, you'll know what to look for in years to come. You may even persuade your neighbors to join you in a program to suppress those invasives. Many invasives first appear in spots where native plants have been damaged or soil has been disturbed, so keep an eye on those sites. Earth-moving activities, floods, and storms can create conditions where invasives can get a foothold or go out of control.

Even if you have masses of invasives, don't give up. Take a base map of your property, mark where different kinds of invasives are growing, and try to judge the level of infestation: light, moderate, or heavy. Decide which stands of invasives need attention right away. Get rid of invasives that are threatening to choke out native plants that are especially valuable to wildlife, such as dogwood shrubs. Heavy infestations of invasives upwind of areas with excellent, so-far-uninvaded habitat should be treated right away. If you find a brand new patch of invasives, get rid of it quickly before the plants spread. For some animals, especially rare ones such as New England cottontails, you may need to manage the invasives themselves to preserve hiding cover, even as you try to change the balance in a habitat patch from predominantly invasive to mainly native plants—a process that can take years or decades.

Techniques for suppressing invasives fall into two categories. *Manual techniques* include pulling up by the roots, cutting, mowing, and girdling; these approaches tend to be cheap but labor intensive. *Chemical techniques* rely on herbicides, which may be needed in areas where invasives have overrun the habitat. Combining manual and chemical treatments can let you gain control over infestations.

Weed wrenches and other similar tools can pull up invasives if the shrubs are not too big. This approach works well when the ground is wet and the soil is soft.
Rachel Stevens

MANUAL TECHNIQUES

Pulling. In spring or at other times when the ground is wet and the soil is soft, you can pull smaller invasives by hand. Multiply your strength and leverage by using a pulling tool (some commercial types are named Weed Wrench, Uprooter, Root Talon, Extractigator, and Puller-bear). Pull slowly to get as much of the plant's roots as possible and to minimize soil disturbance, since many invasives get their start on the bare ground of disturbed sites; you can also replace any disturbed soil and cover the spot with leaves. Pulling probably won't work on larger plants or those with taproots or mature root systems because chunks of roots left in the ground will resprout. A mattock or Pulaski tool can also be used to dig up invasives' roots.

Mowing and Cutting. By mowing or cutting them, you can slow the growth of many invasives and lessen their ability to produce seeds. Be prepared to mow or cut at least twice a year and, to achieve greater control, up to six times annually—and to keep mowing those pesky plants for years. Some species, like Oriental bittersweet, resprout abundantly after cutting, which means you can actually invigorate the plant by cutting it. Another approach is to mow or cut an invasive, which stresses the plant, and then after it spends resources sending up new sprouts and leaves, treat them with herbicide. Trees or shrubs that are too tall to safely spray can also be cut down and herbicide applied to stumps or resprouts. Goats, sheep, and pigs graze down invasives, and prescribed burning kills some invasive plants.

Girdling. Girdling is a directed approach that kills single trees. Cut away a strip of bark all around the trunk, slicing deep enough to remove the cambium, or inner bark, which is the tree's lifeline. You can use an axe, knife, or saw; some people make two parallel cuts around the trunk that are six inches apart, then remove the outer and inner bark between the two cuts using a sharp axe blade or a chisel. Girdling takes less work than cutting and removing a tree; it doesn't require using herbicides, and it results in a standing dead tree that turns into a snag, providing perches for birds. Don't try to girdle ailanthus, also called tree of heaven, because that just makes it resprout heavily.

Tilling. Farmers go over fields with tilling equipment to control weeds, and this approach can be used against invasives in old fields where soils already have been disturbed. Tilling works against annuals and shallow-rooted perennials, but root fragments of invasives with dense root systems, such as tree of heaven, can resprout after tilling. Till during dry periods to make it less likely that root segments will regrow.

Flooding. If you can manipulate the water level of a wetland or a pond, you may be able to kill invasives by flooding them and depriving them of oxygen. This technique has worked on buckthorn, barberry, Oriental bittersweet, multiflora rose, autumn and Russian olive, and tree of heaven. Flooding during the growing season does more harm than flooding before plants leaf out in spring. Check with local and state authorities to make sure you don't violate any wetlands or waterway regulations before trying this approach.

CHEMICALS

Most of us don't like the idea of using herbicides in nature, but there are ways to employ them that minimize hazards to ourselves, non-target plants, and the environment.

Choose the least persistent, least toxic, and least mobile herbicide that can do the job, and, as far as it's possible, one that is selective and only harms your target plant. Rather than indiscriminately spraying, isolate individual invasive plants and attack them. Educate yourself by studying a 2001 publication by The Nature Conservancy that is widely available on the web: *Weed Control Methods Handbook: Tools and Techniques for Use in Natural Areas*, by Mandy Tu, Callie Herd, and John M. Randall. Your state's university extension service probably offers fact sheets, articles, and manuals with good localized information. *Invasive Plants: A Guide to Identification, Impacts, and Control of Common North American Species*, by Sylvan Ramsey Kaufman and Wallace Kaufman, has a chapter, "Managing the Good, the Bad, and the Ugly," with a wealth of practical information.

Different ways of using herbicides include spraying the leaves of targeted plants; cutting a

Backpack sprayers direct herbicide at invasives, limiting damage to nearby native plants. Landowners can use smaller sprayers available from garden stores. Choose the least persistent and least toxic chemical that will do the job.
Judy Wilson, CT DEEP

tree or shrub near ground level and immediately spraying, squirting, or painting herbicide on the exposed stump; and spraying a six- to 12-inch band of herbicide around the circumference of the plant's trunk, about a foot above the ground (this approach, known as basal spraying, works best on young trees with smooth bark; its effectiveness can be boosted by first girdling the tree). For larger trees, use the "hack-and-squirt" technique, in which you hack a series of notches into the trunk, and then squirt a concentrated systemic herbicide into the fresh wounds.

Research suggests that glyphosate, the active ingredient in many herbicides, may kill helpful bacteria in bees' guts, making the insects more prone to infection and death. This may happen when the herbicide is sprayed on target plants' foliage but also gets on flowers that bees visit for nectar. Applying herbicides to cut stumps or stems is probably less apt to harm bees.

Always remember that the best way to avoid having to use pesticides—and needing to spend hours, days, weeks, and years fighting invasives—is to not let those noxious plants take over your land in the first place.

A ROGUES' GALLERY OF INVASIVES

Following are descriptions of some common invasive grasses, herbs (flowering plants), vines, shrubs, and trees, with recommendations on how to fight them.

Japanese Stiltgrass

The lance-shaped lime green leaves are about three inches long, with a shiny silver midrib like a slightly off-center stripe. Japanese stiltgrass seems to have come to North America in the early twentieth century as dried, seed-laden packing material for shipments of Oriental porcelain. Today, it is found from southern New England to the Carolinas and west to Texas and Illinois. Japanese stiltgrass invades land disturbed by grazing, mowing, or logging. It shows up on road margins and along hiking trails in mature forests. It can be spread by logging and mowing equipment and by hikers' boots. Stiltgrass grows one to three feet tall, forming dense, sprawling mats. An annual, it reproduces through its seeds, with each plant producing 100 to 1,000 seeds a year.

For a small area, pull plants out of the ground during damp weather before they flower. Mow or weed-whack larger patches in September before plants produce viable seeds and when it's too late for them to grow back again before the first frost. Pull or mow stiltgrass over several years until all the seeds are gone from the site. For large infestations, use herbicide that kills only grasses and doesn't harm native broadleaf plants.

Johnsongrass

A type of sorghum, Johnsongrass grows mainly in the southeastern and south-central states but also shows up in locales as far north as southern Canada. It has long, thick roots, can grow eight feet tall, and looks like a corn plant with longer, narrower leaves. Johnsongrass reproduces through seeds and rhizomes (roots that spread below ground). A single plant can shed 80,000 windborne seeds a year. Forest edges, disturbed areas, old fields, and streambanks are some of the habitats Johnsongrass can invade. Since its growth produces so much biomass, stands of this tall grass become fire hazards during drought.

Pull out small patches by hand when the soil is wet; try to keep from breaking off parts of the roots so they don't remain behind and resprout. Several years of heavy grazing by hogs or goats can kill this invasive by preventing the plants from seeding and denying nutrients to their roots. A selective herbicide sprayed in consecutive years can eliminate Johnsongrass.

Garlic Mustard

The young leaves of this biennial smell like garlic when crushed. First-year plants look similar to violets; second-year plants have triangular toothed leaves and send up a flower stalk one to three feet tall in mid-spring. The plants die after flowering, and their small black seeds (up to 500 per plant) spread by gravity, water, wildlife, and human activities. Garlic mustard is found from southern Canada to Georgia and west to Wisconsin. It can take over forest edges, riverbanks, and road margins. Garlic mustard often spreads from disturbed areas into undisturbed woodlands, where it chokes out spring wildflowers and tree seedlings; it may also inhibit the growth of underground fungi that other plants need to help them take in nutrients. Butterflies that deposit eggs on native mustard plants sometimes lay their eggs on garlic mustard; after the caterpillars hatch, they don't survive because the chemistry of the garlic mustard foliage differs from that of their normal host plants.

Garlic mustard pulls out of the ground easily when the soil is damp. For large patches, cut the plants close to the ground in spring and then spray resprouts with glyphosate in late fall. If you pull plants that have begun to flower, don't leave them on site, as their seeds may still develop.

Purple Loosestrife

Tall spikes of beautiful purple flowers don't begin to make up for this alien's propensity to invade damp soils along shores of lakes, streams, and wetlands and grow so thickly that other plants can't survive. Dense monocultures of purple loosestrife stand two to seven feet high and represent lost habitats for waterfowl, wading birds,

Purple loosestrife invades damp soil along shores of lakes, streams, and wetlands and grows so thickly that native plants can't survive. Dense monocultures of loosestrife represent lost habitats for wildlife. © *iStock.com/AnnekeDeBlok*

muskrats, frogs, toads, and turtles. Each flower spike produces thousands of tiny seeds that spread by wind and water. The plants develop robust root systems. One of the most successful invasives, purple loosestrife has colonized parts of all Canadian provinces and the lower 48 U.S. states, with the worst infestations in the Northeast. Conservationists estimate that purple loosestrife now dominates half a million acres.

This is an invasive to confront and do battle with right away. Pull or dig up small pockets of loosestrife (try to get all the roots, or the plants will resprout). Large, well-established stands will be very difficult, if not impossible, to eradicate. Two imported beetle species offer hope of future biological control. The herbicide glyphosate kills loosestrife, but it also harms other wetlands plants. Before using herbicides, contact your state's department of natural resources or your local cooperative extension service. Some states certify and license private contractors with the equipment and know-how to eliminate this invader.

Japanese Knotweed

Introduced as an ornamental in the 1800s, Japanese knotweed now infests streambanks, low areas, and forest edges from southern Canada to Louisiana. It has heart-shaped leaves and sends up multiple hollow stems, similar to bamboo, that grow four to 10 feet tall. The stems die over winter, and new ones grow back each spring from an extensive underground root system. Upright flower clusters bloom white in late summer and produce winged seeds that are blown by the wind or carried by water to new areas. The dense growth of Japanese knotweed smothers native plants.

For a small patch, clear away last year's canes in spring and lay down plastic tarps to weaken the plant. Later, take off the tarps, let the plants sprout, and apply herbicide. Glyphosate will kill this invasive; for knotweed growing along a stream or wetlands, use formulations approved for aquatic habitats. You can also let the plants sprout in spring and grow until early or midsummer, and then cut them back and spray regrowth with herbicide in August or early September. Herbicide can also be applied directly to cut stems. Whichever method you use, it can take years to kill the roots of this tough, persistent plant.

Autumn Olive and Russian Olive

These shrubs or small trees grow to 30 feet tall. Both types have silvery foliage; autumn olive has oval leaves, while those of Russian olive are lance-shaped. Both species have thorns. Invasive olives put out small fragrant flowers that, after being pollinated by insects, yield reddish fruits (autumn olive) or yellow fruits (Russian olive) slightly smaller than a wild blueberry. The seeds are dispersed in bird droppings. Autumn olive can be found from Maine to Florida and west to the Great Plains. Russian olive causes problems from Virginia northward.

Pull young seedlings by hand or with specialized tools that multiply force to wrench larger plants out of the ground. The best time for this approach is in spring when soils are damp and loose; resprouts can be hit with herbicides. You can cut down invasive olives with a chainsaw and paint the cut stumps with glyphosate, a treatment that works best in late summer. Spraying the foliage and applying herbicide to the base of the trunk can also kill autumn and Russian olive.

Bush or Shrub Honeysuckles

Three types cause major problems: Amur, Morrow's, and Tatarian (sometimes spelled "Tartarian") honeysuckle. They grow up to 20 feet tall and nine feet wide and are found from New England south to Georgia. Nectar-feeding insects and hummingbirds pollinate these shrubs' flowers, whose colors range from off-white to pink and crimson. The resulting plump, round berries are

Many invasive shrubs, such as bush honeysuckle, leaf out early in spring, making it easy to find and flag them for removal.

red or orange and contain seeds that are spread via bird droppings. Two native bush honeysuckles, swamp fly honeysuckle and American fly honeysuckle, are smaller and live only in wooded areas; invasive honeysuckles prefer more light, growing in forest edges, old fields, and shrublands. Our native honeysuckles have a white pith (the soft, spongy center of the stem); that of invasives is tan or yellow.

Invasive honeysuckles put out pale green foliage before native plants leaf out, making them easy to find in early spring. Pull small honeysuckles out of the ground, either by hand or using a weed-pulling tool. Repeated mowing or lopping (three to four times each growing season) can control small infestations. Spray herbicides on the leaves in late summer, or apply them to cut stumps from July through late November.

Japanese Barberry

Thorny stems and oblong scarlet berries distinguish this invasive shrub. Japanese barberry grows to six feet tall and spreads three to seven feet wide. It is found from Atlantic Canada south to North Carolina and Tennessee. It prefers partial sunlight but can also grow in shade, especially in younger forests. A Japanese barberry patch spreads when the shrubs' roots creep outward and when branches droop to the ground and send down roots. Fallen leaves decompose and change the soil's chemistry so that few native plants can survive beneath Japanese barberry. Birds such as ruffed grouse and wild turkey eat barberry fruits and spread the seeds in their droppings. European barberry, a close relative, is also invasive. A native barberry species, American barberry, grows mainly on rocky slopes in the Appalachians from Pennsylvania south to northern Georgia and should not be destroyed.

Pull out Japanese barberry by the roots in early spring, wearing gloves to protect against the sharp thorns. Spraying herbicides on the leaves or the bark at the base of the stem also controls these shrubs. Propane torches or flame weeders can direct high-intensity flame at the base of the shrub clump, heat-girdling the stems and killing buds at the top of the root crown.

Multiflora Rose

Multiflora rose infests many parts of the East. This rugged shrub has a fountain-shaped appearance, and wicked thorns stud its arching canes. Sprawling patches take over old fields, forest edges, and areas in woodlands where they can get enough light. White flowers develop into red rosehips that hang on the shrubs into winter; birds and mammals eat the hips and scatter their small seeds via their droppings. In a year, a single plant can produce a million seeds, which can survive in the soil for 20 years. The U.S. Department of Agriculture once promoted multiflora rose for "wildlife cover plantings," and this shrub does provide dense cover for small mammals and birds but at the expense of native shrubs that offer better food to wildlife.

Mowing with heavy-duty equipment three to six times a year may subdue this shrub but probably won't kill it. Herbicides can be sprayed on the foliage after flower buds form. You can also draw the canes aside using a pitchfork so that a partner can use loppers or a chainsaw to cut them at their bases; apply the herbicides glyphosate or triclopyr immediately after cutting.

Buckthorn

Two species, common and glossy buckthorn, are widespread in the East from North Carolina and Tennessee on north. Shrubs or small trees up to 20 feet tall and 10 inches in trunk diameter, they form dense stands and crowd out native plants. (Two native buckthorns also occur in the East: Alder-leaved buckthorn is a low shrub rarely taller than three feet, and Carolina buckthorn has red fruits and toothed leaves that are smooth on both sides. Field guides and photos will help you tell them apart from the invasives.) Mature common buckthorns can resemble abandoned apple or plum trees from a distance. Buckthorns' leaves look somewhat like those of dogwoods, with prominent veins. Glossy buckthorn favors wetlands and moist soil; common buckthorn invades woods, woodland edges, utility corridors, and abandoned fields. Both types produce black fruits a quarter-inch in diameter. Mammals and birds eat the fruits and disperse the seeds; the European starling (itself an invasive species) is a major disperser.

Wrench, pull, or spray herbicide on small plants. Cutting and mowing can hold buckthorns in check, and large buckthorns can be killed by cutting them down and applying herbicide to the stumps. In grasslands, prescribed burning will kill buckthorn seedlings and, after several burnings, mature plants as well.

Norway Maple

Planted as urban and backyard trees, Norway maples produce deep shade as well as abundant winged seeds. The tree's tightly ridged bark resembles that of white ash, and the leaves have five lobes edged with sharp points. Norway maples are very shade tolerant, and if these trees establish themselves in deciduous forests, the heavy shade they cast will reduce the diversity of plants at ground level, so that eventually only Norway maple seedlings will grow on the forest floor beneath their progenitors. Norway maples

Shrubs growing beneath power lines provide travel corridors for wildlife. Unfortunately, the shrubs covering this right-of-way are mostly invasive buckthorns.
Jim Oehler

are still sold by nurseries. Landowners should avoid planting them and try to kill or remove them where practicable.

Pull seedlings by hand, and use a weed wrench to lever up small saplings. Basal bark spraying or stem injection of herbicides (also called the "hack-and-squirt" method) can kill larger stems. Fell a Norway maple with a chainsaw, and then paint a systemic herbicide on the exposed cambium tissue (the living inner bark) to kill the tree's roots.

Ailanthus (Tree of Heaven)

Ailanthus is a deciduous tree that, with its large compound leaves, resembles sumac when it's young and walnut and pecan trees when mature. Native to China, ailanthus was brought to our continent as an ornamental in the 1700s and has now overspread the East. It colonizes vacant city lots, rubble piles and cracked pavement in industrial zones, strip mine spoil banks, railroad beds, roadside rock cuts, old fields, and forest edges. Ailanthus needs at least partial sunlight to grow. The roots of established trees send up copious sprouts to form dense colonies. Chemical compounds produced by ailanthus roots and from the tree's decomposing leaves suppress the growth of competing plants. Incredibly prolific seeders, mature female trees can produce 300,000 seeds in a year, each of which consists of a single seed inside a thin paper wing. One ailanthus on the upwind edge of an old field can swamp the field with seedlings in just a few years.

Ailanthus resprouts vigorously when cut down. You can pull up seedlings by hand or with a wrenching tool, but root pieces left behind will quickly resprout. Girdling causes ailanthus to resprout heavily. Using herbicide in one way or another seems to be about the only way to reliably kill tree of heaven. With the "hack-and-squirt" method, you hack notches in the tree's smooth bark, and then squirt a systemic herbicide into the wounds. Cutting trees in summer and painting or spraying herbicide on the cut stumps can be effective. You will probably need to hire a professional to deal with a particularly bad infestation.

Kudzu

Widely planted in the South for shade, erosion control, and livestock forage, kudzu now shows up as far north as Massachusetts, New York, and Michigan. Incredibly fast growing (up to a foot a day), kudzu vines kill or damage shrubs and trees by growing over them and blanketing them with its leaves. Winter's frost can kill kudzu vines, but the plants' roots—they can penetrate 10 feet into the ground—send up new vines in spring. One root yields up to 30 vines, each of which can extend 100 feet or farther. Kudzu grows in fields and along forest and road edges. Kudzu seeds are not particularly viable, and the plant spreads largely through its own growth. Ecologists estimate that kudzu has invaded more than seven million acres, mainly in the Southeast, where mild winters, hot and humid summers, and ample rainfall fuel its growth and spread.

The way to control kudzu is to kill its extensive root system. Cut the vines just above ground level and continue to mow resprouts every month for two growing seasons to deplete carbon stored in the roots. (Be sure to safely destroy all cut material.) Goats and pigs can limit kudzu through their feeding; pigs eat the plants' roots as well as the vines. Kill the roots by cutting the vines and painting a systemic herbicide, such as glyphosate, on the cut stumps.

Oriental Bittersweet

This deciduous woody vine has alternate, oval leaves and small, greenish flowers clustered along its stems. Vines up to 60 feet long climb into trees, wrapping around their trunks and branches. In autumn, bittersweet's showy crimson fruits are wrapped in yellow sheaths. Birds feed on the fruits and disperse the seeds in new areas. Seeds also spread to waste sites when people throw out old floral arrangements. Oriental bittersweet can take root in forest shade, and then quickly respond to gaps caused by storms or tree-cutting and climb upward toward the light. Sometimes called "the kudzu of the North," it ranges from Maine south to Louisiana and west to the Great Plains. Oriental bittersweet can hybridize with a native species, American bittersweet, and may threaten the much rarer native plant's genetic identity, as well as outcompeting and displacing it.

Cut the vines four to five feet above ground and also at ground level so that new sprouts can't climb up the old vines. After cutting, when suckers arise from the roots, spray them with the herbicide triclopyr.

10

Habitat Around Your House

BY ITSELF, A PLAIN GRASS LAWN IS STARK AND VISUALLY UNINTERESTING—which is why most homeowners add plantings such as shrubs, flower beds, and specimen trees. Extend this desire for natural beautification further, and you move into *natural landscaping*, the use of gardening and landscaping practices to add visual and textural diversity to the area immediately around your house, with the excellent benefit of attracting and succoring wildlife.

If you own a small property, can you build a yard, a garden, and an immediate landscape with all of the habitat components that a generalist species, such as an American robin, needs? Definitely. What about a wood thrush? Perhaps, if your property borders a sizable tract of mature woods and if it offers attractive features such as running water and a concentration of large and midstory trees. What about a downy woodpecker, a phoebe, a barred owl, an American toad, a garter snake, a cottontail rabbit, or a red fox? It's possible that you can attract some or all of these animals.

Research has shown that seeing wildlife around your home—hearing birds singing, glimpsing

Cedar waxwings cluster at a bird bath. Adding a water source, planting a butterfly garden, or replacing exotic shrubs with native ones can help wildlife while making your own immediate environment more interesting. © *iStock.com/HealerTeresa*

brightly colored butterflies, seeing a garter snake slither into a stone wall—makes life more interesting and satisfying. It's a visceral human need, to be in touch with wild creatures and to acknowledge that we ourselves are a part of nature. In these increasingly busy and often stressful times, it's soothing to have a space where you can shut out distractions and immerse yourself in nature. And you don't have to visit a national park to do that.

The key to helping wildlife in residential areas is to supply as many of the basic life needs—water, food, and cover—for as many species as you can. More and more homeowners are integrating habitat features with their grass yards, informed and encouraged by university cooperative extension services, organizations such as the Audubon Society and National Wildlife Federation, and forward-thinking scientists like Douglas Tallamy, an entomologist and wildlife ecologist with the University of Delaware. His seminal books, *Bringing Nature Home* (Timber Press, 2007) and *The Living Landscape*, coauthored with Rick Darke (Timber Press, 2014), explain how using native plants can help sustain our wildlife, plus deliver many other benefits, including beautifying a property, saving energy, screening out objectionable sights or sounds, cooling the air, sequestering carbon, recharging groundwater, and improving soil aeration.

One acre is a fairly small parcel: 208 feet on a side, if square. Many building lots are as small as a quarter acre, or about 100 feet per side, if square. Even if you own a small lot, you can still make a difference. Whether it's installing a birdbath, putting in a butterfly garden, or planting a hedge of native shrubs, you can help wildlife and the environment while making your yard more interesting and satisfying for yourself and your family.

Try to find neighborly ways of landscaping for nature. The first step in creating an alternative to a lawn is to gain the approval of your neighbors and perhaps your town or township officials. Find out if your municipality has a "weed ordinance." If so, you may need to file an application for natural landscaping.

It's important to talk to your neighbors and tell them what you're planning to do, and why. Explain the advantages of natural landscaping, which include attracting songbirds and butterflies that everyone can enjoy, eliminating pesticide use on your former lawn, reducing maintenance costs, and cutting back on noise and air pollution from mowing. Real estate values actually rise for properties with carefully planned and beautifully carried out natural landscaping. Share with your neighbors printed information on the topic, likely available from your cooperative extension service. Two excellent publications pertinent for all the eastern states, downloadable and printable from the Penn State Extension website, are *Neighborly Natural Landscaping in Residential Areas* and *Meadows and Prairies: Wildlife-Friendly Alternatives to Lawn*, which are both authored by Margaret C. Brittingham, Ph.D. As Brittingham points out, "The more people who practice natural landscaping, the better for the environment and the more acceptable it becomes."

An excellent way to win your neighbors' approval and possibly enlist them in a community-wide effort to make habitat and link habitat areas through wildlife corridors—is to design your lawn alternative so that it clearly appears to be intentional and doesn't look like some neglected weed patch. Brittingham avers that "people like order, purpose, and tidiness," and she offers key strategies to help your landscape look "tended."

Install borders. Mowed edges or setbacks from your property line can keep taller plants from obstructing sight lines or encroaching on your neighbors' lawns, the sidewalk, or the street. You will want to keep some lawn for your own uses and activities, such as children's play areas or places where you can picnic, have an outdoor

The key to helping wildlife in residential areas is to supply as many of the basic life needs—water, food, and cover—for as many species as you can. More and more homeowners are using natural landscaping to integrate habitat features with their grass yards.
Howard Nuernberger, Penn State

get-together, stroll through your habitat, or efficiently access different habitat areas for maintenance. Put in paths that human visitors can use without having to brush up against taller vegetation that may harbor ticks.

Use curving lines where possible. Our aesthetic sensibilities respond much more favorably to curved lines than to sharp corners and hard, straight lines. In the wild, natural areas tend to have curves and zones of gradation at their edges.

Start small. Begin with a discrete, doable project such as planting a small wildflower garden or bringing back a natural forest understory beneath trees on your lawn. From these core areas, move outward to take in more of your yard. Writes Brittingham: "Expand gradually, imitating nature's processes of gradual succession rather than sudden takeover."

Match the correct plants to your site. Trying to grow sun-loving wildflowers beneath shade trees will be frustrating, probably unsuccessful, and potentially unattractive. Don't shoehorn the wrong kinds of plants into areas where they won't thrive.

Use native plants where possible. Research shows that insects—the key foodstuffs of our birds and their offspring—rarely eat non-native plants since they lack the enzymes to digest such vegetation. Many native plants are just as beautiful as non-natives. That doesn't mean you need to yank out every foreign-sourced plant in your garden. However, you should be aware of and never plant species that might aggressively invade other habitats and displace native vegetation. Being a good neighbor includes making sure your plants don't trespass in your neighbors' gardens.

Manage invasives. Consult your state's list of noxious weeds and invasive plants, and gain information from specialized books or your extension service on how to remove invasives. The Nature Conservancy is an excellent source of information. To discourage invasive plants, a wildflower meadow should be mowed yearly. Hand-pulling, mechanical removal, and targeted trimming can control some species. (See Chapter 9, "A Plague of Invasives," to learn more.)

Humanize your habitat projects. Well-maintained paths, handsome benches, bird feeders, birdbaths, sundials, and well-placed objects such as antique farm tools or sculptures increase the perception that your natural landscape has been planned and is not simply a shabby untended space.

Advertise and educate. Put up special natural landscape-habitat yard signs. If you enroll in the National Wildlife Federation's Garden for Wildlife program and fulfill its requirements, you can purchase attractive, informative signs stating that yours is a certified wildlife habitat. The Xerces Society sells signs for pollinator habitats.

Certify your landscape as a wildlife habitat. The National Wildlife Federation is at the forefront of this movement. You need to demonstrate that your garden or yard provides wildlife with food, water, cover, and places to rear young and that you use sustainable gardening practices. Adding native plants, bird feeders, nesting boxes, or small water features can lead to certification, either from a national entity or a local or state conservation organization.

Share your enthusiasm. Draw up a map of your natural landscape and place copies in a brochure box in your yard. Include a list of the plants you used and where you got them. Host a lawn party for your friends and neighbors and lead them on a tour.

CREATING A SMALL WOODS

Planting and caring for trees on an acre or less is a very manageable undertaking. You can create a compact woodland by buying and planting native hardwood trees and making sure the ground beneath them develops the characteristics of a natural woods, complete with leaf litter, fallen branches, and shade-tolerant shrubs, rather than grass. As they mature, the trees will make shade, provide food and nest sites for birds, gladden the eye with colorful fall foliage, display their bare branches and interesting shapes in winter, and green up prettily in spring.

Consider having a soil test done before deciding which trees to plant. If your soil is acidic (with a pH less than 5), focus on acid-tolerant species. A publication developed by university extension scientists, *The Woods in Your Backyard*, includes tables showing different trees' soil and site requirements, shade tolerance, susceptibility to injurious agents, growth rates and longevity, and food value to wildlife. Elsewhere in the same publication are detailed instructions on how to plant and care for trees and shrubs.

Eight to 10 trees, strategically spaced, will make a nice grove on a one-acre lot. For diversity's sake, plant trees of several different species. Good candidates include oaks, maples, hickories, beech, black cherry, and black gum (tupelo). Consider planting one or more American elms from cultivars that have been bred to resist Dutch elm disease, which has killed many elms throughout the East. (Most hybrid elms are crosses between American, European, and Asian elms that have shown resistance to the disease.) Sycamores are a good choice for rich soil near streams, but on small tracts, their wide-spreading, fast-growing roots may heave up sidewalks and infiltrate water and sewer lines. Avoid ashes because an invasive insect, the emerald ash borer, is attacking and killing ash trees regionwide. Adding several conifers—spruces, firs, or hemlocks—will provide birds with thermal cover in winter. Spring is the best time to plant. You can plant bare-root or containerized seedlings. Or, since you're purchasing and planting a limited number of trees, you may opt to buy larger ones from a local nursery that will more quickly grow to a significant size.

Think about where wildlife habitat already exists on your land and on neighboring properties. Can you connect two or more separate habitat patches through tree planting? Or, by selecting certain kinds of trees, can you offer wildlife a type of food that's scarce in your area? Black gums produce nutritious fruits in autumn that, thanks to the gums' striking orange, red, and yellow foliage, migrating songbirds quickly find and eat. Birches and oaks support large numbers and varieties of insects, which in turn feed many songbirds and their young during the spring and summer breeding season.

Try to create vertical diversity.

At the very top of your woods, *canopy trees* catch sunlight with their leaves, taking in energy that allows the trees to flower prolifically, fix carbon and release oxygen, and produce fruit and seeds. Wildlife that uses the canopy includes birds, squirrels, raccoons, spiders, and insects such as katydids, walking sticks, and caterpillars.

Understory trees are shorter tree species that can tolerate some shade, along with younger specimens of canopy trees. Dogwoods, ironwood, American hop hornbeam, black gum, American holly, umbrella magnolia, pawpaw, and redbud are examples.

In the *shrub layer,* plant attractive shrubs that offer food and cover to wildlife, such as rhododendron, mountain laurel, spicebush (deer avoid eating spicebush leaves), flowering dogwood, arrowwood viburnum, oakleaf hydrangea, azalea, witch hazel, and leatherwood. All of these plants have adaptations that let them thrive in low-light situations, intercepting shafts and flecks of light that make it through the canopy and understory layers.

If your house is in the woods, planting understory trees can help add vertical layers to your home habitat. Black gums are shade-tolerant, medium-sized trees that turn a beautiful color in fall when they produce fruits that birds find irresistible.

A *herbaceous layer* of plants grows from the surface of the ground to about four feet. This zone includes ferns and mosses, flowering plants, seedlings of trees and shrubs, and vines. Some ground-level plants are annuals, growing from seed each year and then dying; others are perennials, coming up from the same roots year after year. A few examples from this diverse layer include trout lily, Virginia bluebell, Dutchman's breeches, squirrel corn, spring beauty, bloodroot, and trillium (all spring ephemerals), maidenhair fern, skunk cabbage (in damp areas), mayapple, wild ginger, bird's foot violet, twinleaf, and woodland phlox. You can buy seeds, bulbs, or transplants of many of these natives either from a local nursery that specializes in native vegetation or from an online seller.

In the *ground layer*, bacteria, fungi, earthworms, and insects break down organic matter. Fallen branches and leaves, bits of bark, twigs, and seeds and seed casings decompose in this shaded, humid environment, their constituent elements gradually recycled through the soil to become new plant growth. Water from rain and melting snow filters into the soil. Springtails, ants, millipedes, snails, salamanders, birds, and small mammals live here.

A BACKYARD MEADOW

Planting a small plot of warm-season grasses and native wildflowers in a sunny area will add beauty and interest to the landscaping around your house while offering food and cover to rabbits, meadow voles, and songbirds. Caterpillars will munch on wildflowers' leaves (and local birds will dine on the caterpillars), butterflies and other pollinating insects will sip nectar from flowers, and birds and small rodents will eat seeds produced by wildflowers and grasses. Audubon maintains a Native Plant Database listing the best species for birds in different areas.

There are two main types of meadow plants: annual and perennial. Annual wildflowers (the type most often found in packaged or canned meadow mixes) grow quickly and provide a splash of color but then need replanting in future years. Perennial wildflowers and warm-season prairie grasses take longer to establish, usually two to three years, but then they will thrive for many years. A perennial meadow will require only occasional weeding, though it should be mowed once each year to keep invasive weeds from becoming a problem. You can also include some annual wildflowers to add color and diversity during the first year or so.

A former lawn will make a good planting site. After laying out your plot, remove as much sod as possible. Dig up the area with a shovel or a rototiller. Dispose of clots of grass, weeds, and roots. If you use a rototiller, till just deep enough to dislodge the current growth; work the soil too deeply and you'll bring more weed seeds to the surface, where they will germinate. Some folks kill weeds and lawn grasses with herbicides; others opt for *solarization*, which involves covering the sod with plastic during summer and letting the sun's rays cook the grass until it dies.

It's fine to prepare your ground and plant seeds in autumn. Wait until after one or more killing frosts to be sure the growing season has ended and the seeds won't sprout until the following spring. For a spring planting, decide during winter which types of grasses and wildflowers to plant and order your seeds. Look for a native wildflower company in your state or region and make sure their seeds are certified weed-free and suited to your climate.

Depending on your latitude, you will probably plant between March and May. Remove any remaining vegetation from your seedbed. Tamp or pack down the soil (kids' feet work well) so that a footprint barely registers. For sites smaller than half an acre, seed by hand. Divide your grass-and-wildflower seed mix into two equal parts, placing each part in its own bucket. Add clean white sand and mix it with the seeds. Take one bucket and hand-broadcast its contents evenly over your planting site. Then do the same with the second bucket while walking in the opposite direction. The sand will make it easier to sow your seeds evenly because it will show up against the soil, letting you know if you have covered your whole planting area.

After seeding, do not cover up the seeds with hay or peat moss or dirt. Press them into the soil by laying down a scrap of plywood and treading or jumping up and down on it. Or simply walk back and forth over the seedbed. (Again, kids' feet can be employed for this task.) An ideal planting depth is about a quarter inch.

Warm-season grasses put most of their energy into developing their root systems during the first growing season when it can be hard to see their above-ground growth. Also, some seeds may remain dormant until the second growing season. So patience is a virtue. As you wait for the grasses to establish themselves, pull out invading weeds. Some people kill emerging weeds by hand-spraying them with herbicides. You can also control weeds by mowing the plot to a height of six to 10 inches, as long as the warm-season grasses are shorter than that. You may have to mow several times, knocking back the weeds until the grasses overtop them.

BUTTERFLY GARDEN

A butterfly garden should have two types of plants: ones whose foliage caterpillars (larval butterflies) can feed on; and one whose flowers provide nectar, sustenance for adult butterflies. It works best if you plant both caterpillar host plants and butterfly nectar plants in clusters.

Caterpillars don't move very far (many will stay on a single plant), and butterflies are attracted to large blocks of color, especially flowers of purple, red, yellow, orange, or pink. Site butterfly gardens where there's ample sunlight and in areas shielded from the direct wind, as butterflies are not strong fliers, and too much wind will keep them from flying. Add a few rocks for butterflies to bask on, and provide areas of wet sand or mud. You can plant butterfly plants in part of your vegetable garden. Avoid butterfly bush (*Buddleia* species), an Asian variety proving to be an invasive species. (For more information, see the section on butterflies in Chapter 12, "Helping Other Wildlife.")

A monarch caterpillar dines on milkweed foliage. Milkweeds are excellent plants for natural landscaping. More than 20 species grow in eastern North America.
© *iStock.com/krisinger1*

HOME WATERS

Draw in birds by giving them sources of water for drinking and bathing. There are two main types of bird baths: those located above the ground (either on a pedestal or suspended from a tree), and those placed directly on the ground. Pedestal baths will attract the same kinds of birds that visit elevated bird feeders, such as chickadees, house finches, and tufted titmice. Ones on the ground will draw in many other birds, since they resemble the kinds of water sources found in nature, including streams and ponds.

You can buy bird baths in garden and pet stores and online. They range from cheap and cheerful to upscale and expensive, and the birds won't care which type you add to your backyard habitat (although you or your neighbors might). What they don't like are slippery surfaces, such as glazed pottery and smooth plastic. Instead, opt for types with rough surfaces, like concrete and terracotta, which birds can get a grip on. Birds like to use shallow water an inch or two deep. A dripper adds motion and sound, both of which will attract birds' attention. You can suspend a plastic jug above a bath; fill the container with water, and add a small hole in the bottom, refilling daily. Site birdbaths near shrubs or tree branches where birds can fly to get away from aerial predators. Changing the water in a bird bath daily prevents mosquito larvae from hatching. Scrub with a brush and mild soap once a week to minimize algae.

Backyard ponds attract insects (dragonflies, damselflies, water striders, and more), birds, frogs, toads, salamanders, and mammals that will drink there and hunt for prey. Ponds can be six feet on a side or larger. You don't want a pond to overwhelm your yard, but you want it to be large enough so that wildlife will use it. A general approach is to decide on a site and then outline a pond using twine or a garden hose; dig out the pond to the desired depth, making sure you have some shallow edges so that birds and amphibians will use the pond (try to keep shallows in shady spots to limit the growth of algae); install underlayment if your area is rocky; spread out a pond liner that overlaps your excavation slightly; cover the bottom with soil, sand, or small rocks; add water slowly; cut away the excess liner to about six inches beyond the pond's margins; cover the edges with soil; and blend it into the landscape by adding native plants.

A male hairy woodpecker sips water from a bird bath. Birds drink and bathe in shallow waters.
Tom Berriman

It's much easier to put in a pond on a level site than on a sloping one. Locating a pond in an area that's usually wet may cause the liner to buckle. (Alternatively, you can probably create a small wetland on such a site.) If you decide to add aquatic plants, they will need some sunlight each day. Partial, intermittent shade from deciduous trees provides cooling in the summer; in spring and fall, when the trees' leaves aren't present, sunlight will warm the water. In and edging your pond, plant deep-water, floating, submerged, and marginal plants. Make sure they are native species and that they are hardy enough to survive in your plant growth zone.

Many books go into detail on how to site and build small ponds. Two concise references for creating ponds in a yard setting are *Water for Wildlife: Bird Baths and Backyard Ponds,* by Jennifer DeCecco and Margaret C. Brittingham, available from the Penn State Extension website, and "Backyard Ponds," a tip sheet from the USDA Natural Resources Conservation Service website. A companion article also available from NRCS is "Backyard Wetland," which gives good basic information on creating features that will hold water for a few days to a week at a time, with suggestions for plantings.

WHAT TO PLANT

When choosing plants and shrubs for your yard, you should select a range of different kinds that will supply food to wildlife year-round. For instance, Juneberries (also known as shadbush and serviceberry) produce fruits from June until August or September, depending on the latitude of the growing site. Blackberries, raspberries, and elderberries fruit in mid-summer. Their fruits will be eaten quickly and avidly by catbirds, cardinals, robins, brown thrashers, summer tanagers, and other birds, along with chipmunks, squirrels, and rabbits. Dogwoods, spicebush, black gum, wild grape, and Virginia creeper put forth fruits later in summer and early fall; neotropical songbirds seek out these high-energy foodstuffs to fuel their autumn migration. In the fall, oaks, hickories, beeches, and pecans produce nuts.

Certain small trees and shrubs produce fruits that have a long "shelf life," with wildlife tending to ignore them until late fall or winter. American

The showy flowering spikes of sumacs produce fruit with a long "shelf life," hanging on well into winter when other foods are scarce. Close to 100 kinds of birds eat sumac fruits.

holly, hawthorn, crabapple, winterberry, and high-bush cranberry belong in this category. Smooth sumac and staghorn sumac offer food to wildlife well into the winter; in addition to providing emergency rations at a time when other foods are scarce, sumacs have showy red fruiting spikes, which add visual interest to a yard or landscape through the cold and often snowy months.

The best guide I have found for learning about and determining which plants to use in a backyard habitat is *The Living Landscape*, by Rick Darke and Doug Tallamy, mentioned earlier. The text is smart and readable. Extensive charts visually show landscape and ecological functions of a huge range of eastern plants, along with interesting notes ("produces nutritious nuts for many mammals"; "supports 235 species of caterpillars"; "provides copious seed for winter birds"). The photographs by Darke will give you ideas on how to blend different groupings of plants and integrate them with features such as retaining walls, water, paths, stone benches, and sculpture.

ODDS AND ENDS

As you develop a home landscape for wildlife, try to sustain and boost diversity in any and all ways. Look closely at your habitat layers from the ground up. Fallen trees provide cover at ground level; some landowners bring in hollow logs and lay them on the ground in strategic places. If you can safely do so, leave dead trees (snags) standing so that woodpeckers can feed on the insects they attract and excavate cavities for nesting. Brush piles and rock piles offer hiding cover and nesting and den sites. You can plant pasture rose or trail a Virginia creeper vine across a brush or rock pile to better blend it into the site.

Tree cavities are wonderful attractions for nesting woodpeckers, chickadees, nuthatches, flying squirrels, and gray squirrels. In suburbs with large old shade trees, barred owls nest in spacious cavities; you just need to provide them with prey-attracting structure in the shrub layer. Adding nest boxes is an excellent way to create artificial cavities in areas where trees are not yet large enough to have developed their own.

Helping Birds

FOREST BIRDS

An amazing variety of birds dwell in forests, from extensive woodlands to smaller woodlots to suburbs built in wooded settings and older neighborhoods shaded with many large trees. Warblers, thrushes, wrens, nuthatches, creepers, vireos, tanagers, towhees, orioles, flycatchers, grosbeaks, woodpeckers, hawks, and owls all use eastern forests. Some of these birds are present only during spring, summer, and fall, as they migrate north to breeding territories, nest and rear young, and then head back to southern wintering areas. Others live in local woodlands year-round.

Some birds nest in extensive tracts of older woods where the canopy (the uppermost branches of trees) is closed, admitting only filtered sunlight. Other species nest in younger forests where the

Red-eyed vireos are among the most abundant birds in mature hardwood forests in the East. They spend most of their time in the uppermost branches of trees and often place their nests near wood roads or forest openings.
Tom Berriman

light is strong and direct and where sapling trees and shrubs crowd together. Birds nest in different ages of woodland and at different heights from the ground level to the canopy. It's the same with feeding, as various species forage in different-aged tracts, at different vertical levels, and on different kinds or parts of trees and other vegetation.

Many forest birds shift from older forest to younger forest at certain times of the year. For example, a pair of cerulean warblers may nest in a tall tree in mature woods at a high elevation; when their eggs hatch, they may fly considerable distances to young forest habitats at lower elevations there to catch insects, which they carry back to their nestlings. Once the young birds develop flight feathers and leave the nest, their parents may lead them to a tract of young forest (a five- to 15-year-old clearcut, for example), where the juveniles learn to feed themselves, protected from hawks by the thick structure of upright stems and overtopping foliage. By eating protein-rich insects, the young ready themselves for migrating to wintering grounds in tropical South America.

As noted earlier, forests that attract the greatest numbers of birds (both individuals and species) have variety in vertical and horizontal structure. They support different types of trees, shrubs, and low plants, offering a range of hard mast (nuts and seeds) and soft mast (fruits and berries). If a wooded habitat has a broad mix of native forest plants, it's likely that there will be at least some mast that birds can feed on each year.

Landowners with large tracts can carefully site timber harvests to add a young forest component and increase diversity both at the forest-stand level and across an entire property. Owners of smaller tracts can boost diversity by planting different types of trees, such as a mix of conifers and hardwoods, or adding food- and cover-producing shrubs in sunny areas. The more vertical layers in your woods—and the greater the variety of trees and shrubs in the horizontal dimension—the more birds you will see.

Another way to keep a forest healthy and attract more and different birds is to control plants that, for one reason or another, come to dominate a habitat and prevent young trees from sprouting and other plants from offering their own unique resources. Common buckthorn, Japanese barberry, multiflora rose, Norway maple, and Japanese stiltgrass are a few of the many foreign plants that invade forest understories. Native plants can sometimes pose similar problems. Hay-scented and New York ferns and American beech are not preferred foods for deer, which browse back other plants; a heavy growth of ferns or beeches makes it hard for the seedlings of other trees (the next generation of forest) to sprout.

Downed woody materials, wolf trees (trees that were left standing in earlier timber harvests because they were crooked or had cavities), snags and stubs, small gaps in the canopy, and clumps of conifers mixed in with hardwood trees all increase forest diversity. You can create some of these features inexpensively. All you need is time, energy, and simple tools such as a chainsaw.

EASTERN BLUEBIRD

A type of thrush, the eastern bluebird breeds across eastern North America. Bluebirds in the South may stay in their home areas year-round, while those in northern areas shift southward in winter. Bluebirds thrive in partially open habitats such as orchards, pastures, fencerows, cutover or burned areas, forest clearings and edges, open woodlots, and suburban gardens and parks.

Perched on fence posts, power lines, or snags, bluebirds scan the nearby terrain for prey, nabbing many insects on the ground and taking some out of the air and from foliage. They eat fruits, especially in winter, including those of dogwood, sumac, red cedar, bayberry, Virginia creeper, black cherry, American holly, hackberry, and elderberry. Outside of their breeding season, bluebirds move around the landscape in small flocks.

LANDOWNER STORY

A Return to Birdsong and Wildflowers

IN 1995, AN HEIR OF MARGERY BOYD gave the Litchfield Hills Audubon Society (LHAS) a 102-acre property in the town of Litchfield in western Connecticut. Twin Brook Farm was graced with meadows, thickets, vernal pools, rock outcroppings, and woods—plenty of woods. The Society renamed the tract the Boyd Woods Audubon Sanctuary. One of three Litchfield Hills Audubon sanctuaries, it's now a popular destination for hikers and wildlife-watchers.

LHAS is a local chapter of the National Audubon Society. Members Debbie Martin and her husband Richard help manage Boyd Woods. Debbie explained that Margery Boyd lived on her farm from 1926 until 1992. An avid birder, she kept daily records of the birds she saw. Said Martin, "Margery's birding diary shows that many kinds of birds that need shrubland and young forest were common during the period when her property was reverting from farmland to forest."

By the time LHAS was given the land, it was 90 percent wooded.

"The woods were beautiful but very quiet," Martin said. "It was obvious that as middle-aged and mature trees had taken over, many species of birds that Margery recorded had disappeared." To add diversity to the landscape, in 2005 LHAS had a five-acre clearcut done, a project supported by funding from the USDA Natural Resources Conservation Service, or NRCS.

"Half of the cut area was allowed to grow back in shrubs," Martin said, "and the other half was planted in conifers. Before long, a variety of birds found and occupied the cut-over area. Folks started hearing and seeing chestnut-sided warblers, blue-winged warblers, eastern towhees, field sparrows, and many other birds that need thick habitat."

Cottontail rabbits were also common. Genetic testing of fecal pellets collected on the sanctuary revealed that they'd been deposited by eastern cottontails, a species characteristic of brushy habitat but one that is not native to Connecticut.

In 2012, when LHAS was approached about creating habitat for the New England cottontail—the region's native rabbit, and a species much rarer than the eastern cottontail—many chapter members strongly objected, mainly on aesthetic grounds. Said Martin, "We'd heard that a clearcut of 25 acres or larger was required" to help New England cottontails, which need large blocks of young forest and shrubland. "After visiting similar large clearcuts in neighboring towns, we were horrified by what we saw: treetops, logs, and huge piles of brush left strewn all over the place. Boyd Woods was a lovely, peaceful spot. We didn't want that kind of a mess on our property."

However, members began to see things differently the more they talked to biologists and foresters with NRCS and the Connecticut Department of Energy and Environmental Protection.

A hiker checks out a patch of aspen three years after a timber harvest set back a forest stand to a younger growth stage. Audubon members wanted to attract New England cottontails to Boyd Woods Sanctuary and bring back birds that need young forest and shrubland. *Richard Martin*

"We learned about the threatened New England cottontail and about all the other species of wildlife that are also struggling due to the disappearance of young forest environments," Martin recalled. Connecticut has identified more than 50 "species of greatest conservation need" that require young forest or shrubland. "On that list were many of the birds that Margery Boyd once counted as common. We realized, then, that we could help bring them back. As an Audubon Society, committed to managing our sanctuary for the preservation of wildlife, how could we not participate in this project?" Martin added: "A turning point came when we were told that we could cut as little as 10 to 15 acres, and not 25. Suddenly, we couldn't wait to get started."

In 2014, LHAS had a logger clearcut eight acres to the west of the shrubland-and-conifer habitat that had been created in 2005. In the winter of 2015, four more acres were cut to the east. Workers built three large brush piles on each logged acre. To continuously maintain early successional habitat for years to come, the chapter committed to managing shrubby areas through periodic mowing and controlling non-native invasive shrubs. Said Martin, "During this process, we have learned so much about the importance of land management practices, and we witness the benefits of this work with each passing season."

Martin said it's now impossible to think of the cut areas as "damaged" or "devastated" by the management actions that the Society has taken. After five years' growth, new young trees—aspens, birch, maple, hickory, and oak—average around 15 feet tall. "Although they still appear somewhat messy," she said, "these new habitats are full of life. Towhees and fox sparrows sing from the brush piles. Indigo buntings, field sparrows, and catbirds join the chorus along the edges of the cut areas. On our annual LHAS Evening Woodcock Walks, an amazing number of American woodcock sing and perform courtship flights over the recently expanded openings."

Where once stood dark forest, renewed sunshine spurs the growth of plants and wildflowers that previously weren't present. "Many of the flowering plants are very beneficial to butterflies and bees," Martin said. "I think the most exciting thing I've found in the clearcuts was a wood lily. I was overjoyed to see these beautiful blooms—one of my favorites, and I probably hadn't seen one in 40 years." She added, "In autumn, many plants go to seed or produce berries, yielding excellent food for wildlife.

"We see lots of garter snakes and a milk snake on one of the paths. The rabbits seem to go back and forth from the extremely brushy areas to the stone walls, where they hide in holes. In the winter, we see coyote and bobcat tracks. Raccoons also

Indigo buntings use edges, thickets, and areas of dense young trees, where the bright blue males sing from exposed perches. Indigo buntings have come back to Boyd Woods following logging. © *iStock/johnandersonphoto*

pass through the clearcuts as well as foxes. There are deer, plenty of squirrels, and last week we saw an opossum wandering around. Some animals, including eastern cottontails, burrow into the brush piles."

New England cottontails have been confirmed on another preserve only three miles from Boyd Woods Sanctuary. "We're confident that when New England cottontails show up here, they, too, will find this habitat accommodating," Martin said.

She added: "Boyd Woods is no longer just a pretty, tranquil place in which to take a walk. As we've created this mix of conifers, shrubland, and young forest habitat—with a total of 17 acres cut between 2005 and 2015—we've truly become a wildlife sanctuary."

Martin has noticed a ripple effect of the habitat work at Boyd Woods. She reported in winter 2018: "Just this week, Rich and I visited a 10-acre clearcut that's currently being done in a nearby town. The landowners had been to Boyd Woods several times, liked what they saw, and this influenced their decision to do a cut of their own."

Margery Boyd had always wanted her land to be used for education and the enjoyment of nature. "Today, we have a perfect opportunity to fulfill her wishes in the promising new habitat that's growing on her old farm," Martin said. "As we watch Margery's birds return, and await the arrival of New England cottontails, we readily admit that making the 'messy' clearcuts that now diversify our landscape was the best thing we could have done."

In 2015, the New England Chapter of The Wildlife Society awarded certificates of recognition to Debbie and Richard Martin and John Baker, of Litchfield Hills Audubon Society, for their commitment to making habitat for wildlife and inspiring others to make similar choices.

The bluebird is the only eastern thrush that nests in tree cavities. People putting up bluebird nesting boxes have helped bring back healthy bluebird populations in many areas. Building one or more bluebird boxes is easy and fun; plans abound on the internet and in books and magazine articles. A bluebird house can be built inexpensively out of scrap lumber or a single one-by-six-inch board six feet long. Drill an entry hole one and a half to one and three-quarter inches in diameter, or leave a long slot running horizontally above the front board of the house (a slot entry also provides ventilation so the box doesn't get too hot in direct sunlight). One side of the box can be designed to pivot on a nail, or the top can be made removable; that lets you clean out old nesting material in late winter before bluebirds start to breed and during summer after a pair has fledged a brood, making it more likely that the birds will use the box for a second nesting. Site boxes on poles with predator guards or on trees four to six feet above ground facing north or east. If possible, orient the box toward a shrub or small tree so that young birds will have a place to alight when they fly from the box for the first time.

Years ago, a retiree of my acquaintance, Melvin Lane, of Alexandria, Pennsylvania, set up a "bluebird route" that at one point was 163 miles long and had 250 nesting boxes. One year (after he had shortened his route somewhat), he had 114 pairs of bluebirds nest in 146 boxes. When starlings and English sparrows usurped the bluebird boxes, they were summarily evicted. Mr. Lane also reported swallows, titmice, flying squirrels, red squirrels, and deer mice using the boxes for nesting or shelter.

AMERICAN WOODCOCK

Also known as the "timberdoodle," the American woodcock lives in young upland forest, brushy woods near rivers and streams, shrublands, and old fields. An adult woodcock is about the size of a robin but with a plumper build. Using its long bill, a woodcock probes the soil for the worms and other invertebrates that make up most of its diet. Woodcock breed across eastern North America from Atlantic Canada to the Great Lakes states and migrate south to spend the winter in lowlands of the Gulf Coast and Atlantic Coast states. Each spring, after woodcock return to their northern breeding grounds, conservationists conduct a range-wide Singing-Ground Survey in which observers drive along specific routes and stop periodically to listen for and count singing male woodcock. Those numbers help the U.S. Fish and Wildlife Service estimate the total woodcock population each year.

From the 1960s to the present, woodcock numbers have fallen by about 1 percent annually—although that trend may have been arrested over the last five years, as many eastern wildlife agencies, as well as private landowners, have made a concerted effort to create and renew habitat for woodcock. In general, though, the woodcock population has declined because the amount of habitat available to these birds has shrunk, in part from human-caused development but mainly because eastern forests have gradually and inexorably matured. As areas of abandoned farmland and younger forest grow to become middle-aged woods, less sunlight reaches the ground and the amount of thick, low-level cover lessens until woodcock no longer use those habitats. Most forests become unsuitable for woodcock when trees reach about 20 years of age.

A tracked "Brontosaurus" machine chewed down old shrubs and small trees, while sparing apple trees, to produce this patch of habitat for woodcock, ruffed grouse, and songbirds.

Woodcock use four slightly different types of habitat.

Courtship and breeding grounds. To attract mates in spring, male woodcock call and make spectacular dawn-and-dusk courtship flights, taking off from clearings, log landings, old fields, pasture margins, and road edges. Landowners can create singing grounds by making forest openings and mowing and brush-hogging old fields. After timber harvests, log landings and skid roads can be planted in grasses and kept open using the same methods.

Nesting and brood-rearing cover. Female woodcock nest on the ground in relatively young hardwood stands near the singing grounds where they bred. Good nesting and brood-rearing habitat has a protective overhead cover of dense hardwood sprouts or saplings growing on soils with good numbers of earthworms.

Feeding areas. Woodcock probe for worms in moist soil under small trees and shrubs. They use alder shrubs along streams, rivers, and wetlands; old fields growing up with hawthorns, dogwoods, and other shrubs; old overgrown orchards; and clearcuts less than 20 years old where shoots and seedlings of aspen, birch, and other hardwood trees grow densely.

Roosting areas. Starting around mid-July, woodcock leave their daytime feeding habitats at dusk and fly to places where they spend the night resting on the ground. These semi-open sites with sparse plant growth include weedy fields, lightly grazed pastures with scattered weeds or brambles, recently logged woods, the edges of old gravel pits, and blueberry barrens. Groundcover should be spotty and open enough to let woodcock detect and escape ground predators such as foxes and weasels while offering some overhead protection from owls. Landowners can make temporary roosting habitats by mowing strips in old fields or hayfields.

Even if you can't make all four woodcock habitat types on your land, you can attract and help these fascinating birds by creating one or more of the cover types they use. Look around and see what kinds of woodcock habitat exist near your property, then try to supply one or two types that are missing or in short supply.

Folks who own woodlands at low elevations can create an ongoing source of woodcock habitat by having a forester draw up a management plan specifying logging on different management units every five to 10 years. Harvesting timber causes stumps and root systems of logged trees to send up thousands of shoots, giving rise to thickets. Woodcock will use newly logged land for singing grounds and roosting habitat and find different ages of regenerating woods for feeding and nesting.

Machines with high-speed mulching heads can chew down old, overmature shrubs that have become too "leggy" and open to remain as good habitat; after being cut during winter, the shrubs will grow back densely in spring. Large "Brontosaurus" cutting machines, with tracks instead of wheels, can work in damp areas without compacting the soil; they can mulch fairly large trees with trunks up to about six inches in diameter and return a habitat area to an earlier growth stage. Contractors in some regions own such machines, and landowners can hire them to create and refresh woodcock habitat, though such projects can be expensive since there's no income from harvesting timber to offset costs. State and federal agencies and wildlife organizations such as the Ruffed Grouse Society may pay for such projects. (Ask your forester for contacts.) A large machine can mulch up to two acres per day. A landowner can do the same thing more gradually using a chainsaw. If you hunt woodcock, grouse, or deer and want to be able to walk through an area managed in this fashion, try to cut the trees so they all fall in the same direction, leaving pathways between the fallen trunks.

Making habitat for woodcock helps fairly common animals such as deer, moose, eastern cottontail rabbit, snowshoe hare, bobcat, wild turkey, and ruffed grouse. Young forest and shrubland also provide homes for many species whose numbers have been falling along with those of the woodcock, including box and wood turtles, golden-winged warbler, whip-poor-will, alder and willow flycatchers, indigo bunting, brown thrasher, New England cottontail, and Appalachian cottontail.

RUFFED GROUSE

In the East, the range of the ruffed grouse extends from southern Canada south through the Appalachian Mountains to northern Georgia; grouse also inhabit the Great Lakes states. The explosive take-off, or "flush," of this chunky, chicken-sized bird has startled many hikers. Hunters avidly pursue grouse, sometimes using flushing or pointing dogs. Ruffed grouse often live in the same habitats as woodcock: young, regrowing forests seven to 20 years old. Unlike woodcock, which migrate south in winter, grouse remain in their home territories year-round, causing them to have slightly different habitat needs. A typical home territory is about 30 acres, although grouse will sometimes move farther when searching for food and cover.

Grouse spend most of their time on the ground, where they feed on insects, plants, fruits, seeds, and nuts, including ones as large as small acorns. They eat apples from old abandoned trees. Grouse fly up into trees, especially aspens, to eat highly nutritious buds.

A habitat will be better for grouse if it has a generous amount of dead wood lying on the ground. To attract females, male grouse "drum" in springtime by perching on downed logs and rapidly beating their wings. Hens nest on the ground, often against the side of a log or stump or among tree limbs knocked down in a storm or left after logging.

Hens lay eight to a dozen eggs. After the chicks hatch, their mother leads them to grassy and weedy areas where the young can find insects—high-protein foods that help them grow quickly. Clearings on the edges of forests and old log landings planted with grasses and wildflowers provide feeding areas. Stands of young hardwood trees growing back following timber harvests or natural disturbances offer excellent habitat. For winter cover, grouse use grapevine and greenbrier tangles, dense shrubs, mountain laurel thickets, blowdowns, and conifers, including spruce, hemlock, and fir. All of these habitat types should exist within 30 to 40 acres.

A friend of mine here in Vermont is an avid grouse hunter. He manages his 200-acre, largely forested property so that it "grows" abundant grouse almost every year. He works to create three major habitat types: dense young trees and brushy woods that provide food and cover in summer and fall; slightly older hardwood stands where male grouse can drum in spring, where hens can site their nests, and where birds can find food in fall, winter, and spring; and clusters of dense, low-growing evergreens to protect grouse from predators, hard rain, and cold. Years ago, my friend had a forester prepare a management plan that included a commercial timbering rotation, so that a significant percentage of his forest stands are always less than 15 years old. My friend "daylights" apple trees (cuts down taller trees that otherwise would shade out the apple trees) and prunes and fertilizes them to improve their health and boost their fruit production. He also mows a network of grassy woods roads and small clearings where hens and their broods can find insects.

My friend has some aspen on his land, which he logs at regular intervals to keep different-aged stands present as part of the habitat mix. The large flower buds of mature male aspens provide winter food for grouse, and dense stands of young regrowing aspen offer daytime loafing and resting cover.

A landowner can manage aspen as an ongoing source of both younger and older trees by cutting one-quarter of the aspen in a stand. (Mature aspens yield valuable wood for products such as fiberboard and paper pulp.) Aspen should always be harvested in winter, which is after the trees have dropped their leaves and stored their energy in their extensive, wide-spreading root systems. Unless the trees are too old and have lost vigor, logging will not kill aspens. Following a winter timber harvest, the trees' roots will send up thousands of new shoots in the spring, quickly regenerating a dense habitat. If a landowner cuts another quarter of his or her aspen every 10 years, after 40 years, the forest will have four separate stands aged 10, 20, 30, and 40 years—a range of ages that grouse will use at different times of the year.

Native shrubs are especially valuable to grouse as a source of fruits, nuts, and browse. Inspect grouse habitats for invasive shrubs such as honeysuckle, barberry, and autumn olive and eliminate or reduce them. Plant native shrubs such as alder, hawthorn, blackberry, raspberry, gray dogwood, pagoda dogwood, and viburnums. (You may need to fence planted or transplanted shrubs to protect them from deer browsing.) If your land already has such shrubs and they're being shaded out by trees, cut down some trees to let in sunlight and restore the shrubs' health. Cut trees can become firewood or left to provide drumming logs; as they slowly decompose, they will offer habitat to many small creatures such as insects and amphibians. Planting a half dozen clumps of shade-tolerant evergreens per acre provides important wintering habitat for grouse.

Recently, the West Nile virus has been found in ruffed grouse in Pennsylvania and Michigan and may be afflicting grouse in other states as well. Research conducted on wild grouse in Pennsylvania suggests that grouse living in high-quality habitats are less likely to die from this mosquito-borne disease than grouse living in habitats of marginal quality.

The Ruffed Grouse Society employs biologists who work directly with private landowners to improve lands for grouse, woodcock, and other wildlife that need similar habitats. RGS offers guidelines on how to manage lands for grouse, and they fund habitat-creation projects on both public and private land. For more information, visit the organization's website.

WILD TURKEY

In some suburban areas, wild turkeys have become so numerous—and so accustomed to and even aggressive toward humans—that they're considered pests. However, in rural areas, they remain wary and popular game birds, hunted in spring (when only males, or gobblers, can be taken) and fall. Many landowners like hearing gobblers' strident, dramatic calling in springtime and enjoy seeing turkeys feeding in newly mown hayfields and shrublands or flying into or out of the trees in which they roost at night.

Turkeys live in forests and farm woodlots with mast-producing trees, and in both deciduous and mixed hardwood and softwood stands. After mating in spring, hens nest on the ground in thick cover such as shrublands, old orchards growing up with small trees, and logged tracts where trees are growing back. Often a hen will tuck her nest in against the base of a log, shrub, or stump. A clutch can be as large as a dozen eggs. After the poults hatch, they feed on insects that they catch in fields, on woods roads, and in forest clearings grown up with grasses and weeds. Creating clearings in the woods as small as a quarter acre can make a habitat more attractive to turkeys.

Because turkeys range widely, even if you have several hundred acres, you may not be able to keep the same birds on your land year-round. However, you can provide food and cover that will cause them to visit your property often and in different seasons. If you own a hardwood forest, manage it to favor mature food-producing trees such as oaks (acorns are a preferred food), hickory, black cherry, beech, and ash. Encourage key food trees by removing trees whose fruits or seeds are not as important to turkeys. Another approach is to cut some overstory trees—perhaps through a commercial timber harvest—to provide more light for grapevines and food-producing shrubs such as dogwoods, viburnums, and hawthorns. You can also plant apple trees, mountain ash, and native shrubs in clearings and sunny spots.

In regions where snow builds up in winter, spring seeps—places where water emerges from the ground—are important habitat features for turkeys. Because the temperature of the water stays above freezing, the seeps remain open and snow-free and offer vegetation and insect foods even in the depths of winter. You can thin the tree canopy around a seep to encourage more green vegetation, being careful not to cut down any valuable food-producing trees or shrubs or blocking turkeys' access to the seep.

NORTHERN BOBWHITE

The bobwhite quail's range extends from the Midwest, Pennsylvania, and southern New England south to the Gulf Coast. Bobwhites are permanent residents wherever they live; short-range fliers, they do not migrate. Range-wide, quail populations have fallen by 80 percent since the 1960s; in northern areas, native bobwhites may be essentially extirpated. The probable reason for this population freefall is a loss of habitat.

Quail love edges. They thrive in open pine forests, fallow fields, and grasslands. Ideal habitat is

Controlled burning next to a pine plantation will spur the growth of edge plants that produce food and cover for quail, such as ragweed, beggarweed, and wild sweet pea. Wildlife agencies in many southern states sponsor landowner habitat programs aimed at increasing bobwhite populations.

a small farm with brushy field-and-woods edges, fencerows, windbreaks, and cropland—a mix of food sources near escape and wintering cover. In recent decades, small farming operations have dwindled as agriculture has become more mechanized, with larger fields and bigger machines; also, development has taken over many old farms. After decades of people suppressing fire, undergrowth has choked out areas that quail once used for feeding, nesting, and resting. Moreover, native plants have been outcompeted by invasive species that provide inferior food and cover for quail and other wildlife.

Quail need foraging areas, nesting cover, and brood-rearing habitat. Juvenile birds feed mainly on insects; adults eat some insects plus seeds, waste grains, green plants, fruits, and berries. Quail have relatively weak feet and are not good at scratching up their food, so they prefer bare or lightly vegetated ground for foraging. Chicks need overhead cover to protect them from aerial predators. A mix of native grasses (also called bunch grasses), broadleaf plants, and native shrubs will provide seeds, insects, hiding cover, and protection from the elements.

Farmers can leave outer rows or corners of crop fields unharvested: corn, wheat, oats, buckwheat, sorghum, millet, rye, and clover. They can refrain from using herbicides on native seed-producing plants such as foxtails, ragweed, beggarweed, and wild sweet pea. Owners of forested land in quail areas can manage their timber to help bobwhites. Periodically burning off the groundcover beneath longleaf, loblolly, and slash pines will boost quail food plants and keep pine stands from becoming so thick that sunlight can't reach the ground. A combination of prescribed burning, disking, and livestock grazing can maintain the correct density of vegetation while encouraging plants that benefit bobwhites. Many southern states have landowner habitat programs aimed at boosting bobwhite numbers.

OWLS

Owls have huge eyes with excellent light-gathering ability, and their hearing is acute; input from both senses lets them locate prey. Their feathers are soft, dampening the sound made when an owl flies. In combination, these traits make owls extremely efficient predators in dim light and darkness. Owls are carnivores, with different-sized owls eating animals that range in size from insects, frogs, and small birds up to rabbits, ducks, opossums, feral cats, and skunks. Rodents are a mainstay in the diet of most owls.

Owls use open, wooded, and riparian areas along with shrublands and wetlands. Anything a landowner can do to increase the number of small animals in a habitat will boost its prey base and likely attract owls—practices such as planting or encouraging seed- and fruit-producing native grasses, wildflowers, and shrubs. Like all wildlife, owls need cover as well as food. Most rest during the day in the dense foliage of evergreen and hardwood trees and in tree cavities; some roost in old barns and abandoned buildings. One species nests on the ground, but most owls site their nests in natural cavities in trees and in abandoned woodpecker holes. Some will nest in artificial boxes.

Listening to owls' dramatic nighttime calling is always a thrill and a great way to get young people interested in nature and the outdoors. An excellent book about owls, which covers their feeding habits and habitat needs in detail, is *The Peterson Reference Guide to Owls* by Scott Weidensaul (Houghton Mifflin Harcourt, 2015).

Barn Owl: This pale-colored, medium-sized owl hunts grasslands, open and partly wooded lowland habitats, and marshes. Barn owls may live in towns and cities if enough good foraging habitat exists nearby. They nest in barns and vacant buildings and use nest boxes (plans are available online) attached to trees, the sides of buildings, or beams or framing members inside silos and barns.

Short-Eared Owl: This uncommon owl is found in prairies, grasslands, wet pastures, and marshes. A short-eared owl hunts by flying low over the ground and hovering in place before

Owls don't excavate tree cavities, but many kinds—including barred owls—use them for resting and nesting. Barred owls will also use artificial nest boxes fixed to trees 12 to 15 feet above the ground. © *iStock.com/Lynn_Bystrom*

WILDLIFE SKETCH

Tyrants Rule

Kingbirds sit on prominent perches and watch for insects to fly past, then sally forth and nab them in midair. The kingbird gets its name because it "rules over" many other birds, aggressively driving them away from the kingbird's own territory.
Tom Berriman

SUDDEN SQUAWKS MADE ME LOOK UP. A raven flapped past, trying to escape a much smaller bird that kept darting in from above and pecking at its back. Letting out loud and aggrieved cries, the raven fled. Its assailant flew up into the air and then came swooping and stunting down. It landed neatly on a dead branch in a crabapple tree in my backyard, settled its plumage, and let out a string of buzzing, stuttering notes. It was a kingbird, and it looked quite pleased with itself for what it had just done.

The eastern kingbird is a member of the Tyrant Flycatcher family. Its relatives include the wood peewee, willow and alder flycatchers, eastern phoebe, and great-crested flycatcher. Birds generally defend their breeding territories, and the kingbird seems to take such behavior to an extreme. Kingbirds fly at and attack potential nest robbers such as ravens, crows, and jays, and even strafe unsuspecting birdwatchers who get too close. They aggressively drive away other kingbirds. Females sometimes even harass their mates.

A kingbird has white underparts, a slate-gray back, and a coal-gray head. A neat white band tips its tail. The crown of a kingbird's head has a bright red patch that usually remains hidden—until the angry bird erects its head feathers and lets that scarlet stripe blaze forth. A kingbird's mouth has a startling red interior, revealed when the bird opens its bill to scold.

Kingbirds perch in open terrain near shrublands or field edges and sally forth to nab insects out of the air. Noisy and obvious, kingbirds are easy to observe: Watch one fly into the wind, hold itself in position on fluttering wings, then dart forward nimbly to snatch a fly or a dragonfly from midair. Kingbirds eat small prey on the wing and take larger prey back to their perches. To subdue a large insect, such as a grasshopper, the bird may bash its victim against a branch before swallowing it. Kingbird fare also includes bees, wasps, beetles, butterflies, and moths. Anything a landowner can do to increase insect numbers will make the local kingbirds as happy as they're going to get.

While they can be irascible on their northern breeding range, on their South American wintering grounds kingbirds reform themselves: They become gregarious, hanging out in flocks and feeding mainly on berries. Kingbirds also eat berries and other fruits toward the end of the breeding season in the north, and while migrating back and forth between their southern and northern ranges.

Kingbirds, especially males, show "philopatry," which means they tend to come back to the previous year's breeding site. I look for them in May, around the time insect populations start to build up—insects that will feed the arriving kingbirds and their nestlings once they hatch. As pretty as they are, these birds deserve their reputation as tyrants. The ones who nest at the edge of my yard definitely rule.

dropping to snatch a vole, mouse, shrew, or other prey. Short-eared owls nest on the ground in small depressions, often beneath a shrub or sedge clump that partly hides the nest. They sometimes nest in grasslands planted to reclaim strip-mined sites.

Great Horned Owl: The classic "hoot owl" and the East's largest and most powerful owl, the great horned lives in forests, woodlots interspersed with farmland, suburbs, and even cities and towns. Its broad range of prey includes rats, mice, voles, rabbits, hares, squirrels, skunks, feral cats, grouse, woodcock, ducks, and smaller owls and hawks. Great horned owls nest in old crow, hawk, and heron nests, in tree cavities, and on rock ledges. Making several-acre openings in wooded areas can boost the amount of prey available to these aggressive hunters. Preserve large old trees that may have sizable cavities in which great horned owls can nest.

Barred Owl: These owls like forests with plenty of mature trees. They sometimes live in older suburban neighborhoods whose shade trees have gotten big enough to develop cavities. Barred owls often live near water, in bottomland woods, and on the edges of swamps and marshes. They hunt in woods, forest openings, old fields, and wetlands; in older suburbs, they look for prey as they fly between the large, widely spaced trees. They eat rodents, flying squirrels (of which they are a major predator), birds, fish, crayfish, and insects. Too big to nest in most woodpecker holes, barred owls seek out natural cavities in large hardwood trees. They also use artificial nest boxes placed 12 to 15 feet above the ground.

Long-Eared Owl: This crow-sized owl often nests among dense evergreen trees, in a cavity, or on an old crow, raven, or hawk nest. Long-eared owls eat small mammals, including voles, mice, shrews, and bats, as well as small birds and snakes. An ideal habitat is a mix of thick conifer stands for nesting and roosting, with nearby grasslands for finding prey.

Eastern Screech Owl: Screech owls live throughout the East from about the middle of Vermont and New Hampshire on south. Adults can be either gray or ruddy in color. Forests, woodlots, old orchards, city parks, cemeteries, towns, and shady suburbs offer a home to this robin-sized owl. Extensive forested areas are less attractive than woods broken up by clearings and fields. Screech owls eat insects, frogs and toads, small snakes and rodents, and numerous birds. They nest in cavities in trees whose trunks (or branches) are larger than about eight inches in diameter at cavity height and in artificial nesting boxes. In late summer, listen for the melodic tremolo call (more of a whinny than a screech) of parent owls trying to keep newly fledged owlets with the family group.

Saw-Whet Owl: The East's smallest owl, the saw-whet inhabits moist woodlands with a dense undergrowth of conifers or shrubs. Small rodents, young squirrels, small birds, and large insects are favored prey. Saw-whet owls nest in woodpecker holes (especially those of flickers and pileated woodpeckers), tree cavities, and artificial nest boxes.

FLYCATCHERS

As their name suggests, these birds are dedicated insectivores: eaters of huge numbers of insects, especially (and in some species almost exclusively) flying insects. In a way, it's difficult to think of flycatchers as "North American birds" because the flycatcher family—*Tyrannidae*, the tyrant flycatchers—is far more varied and abundant in the insect-rich forests of South America, where more than 400 species exist. Of the 10 species that breed in eastern North America, only one, the eastern phoebe, regularly winters on our continent. The others migrate to Central and South America.

Even veteran birders find some of the North American species hard to identify since they tend to be fairly drab in color and uniform in appearance. The following species breed east of the Mississippi River: olive-sided flycatcher (far northern areas), eastern wood pewee (most of the East), yellow-bellied flycatcher (far north), Acadian flycatcher (from the Mid-Atlantic south), alder flycatcher (from Pennsylvania north), willow flycatcher (Appalachian mountains northward), least flycatcher (Appalachians northward), eastern phoebe (northern Georgia northward; winters in the Southeast, migrates north very early in the spring, and heads south later in autumn than other flycatchers), great crested flycatcher (throughout the East), and eastern kingbird (much of North America, including all of the East).

Typically, a flycatcher sits on a branch or other perch and watches for an insect to fly past, then dashes out in swift, agile flight and catches its prey in midair using its bill. Some eastern flycatchers also flutter above vegetation or the ground, then dip down and snatch prey. Flycatchers eat beetles, wasps, winged ants, bees, grasshoppers, flies, leafhoppers, honeybees, caterpillars, moths, katydids, tree crickets, spiders, and others.

Most flycatchers build cup-shaped nests in the branches of trees and shrubs. Yellow-bellied flycatchers sometimes nest on the ground, and eastern phoebes build their nests on small flat surfaces projecting from rock outcroppings, under bridges, and on beams in barns, porches, and other buildings. The great crested flycatcher, our region's largest flycatcher, nests in tree cavities and has the intriguing habit of often adding shedded snake skins to its nest.

Flycatchers use a variety of habitats. Some, such as the great crested and the eastern wood pewee, breed in forest interiors and feed along the edges of clearings where insects abound. Alder flycatchers and willow flycatchers like brushy areas and thickets, especially those in wetlands. Eastern kingbirds and phoebes live in semi-open country, including farms, savannas, shrublands, and woods edges.

All of the flycatchers will be more abundant in places where insects are plentiful. Gaps in mature forest, including both small and large clearcuts, can boost insect numbers. Practices that encourage diverse kinds of lower plants and shrubs—such as thinning a woodland, rejuvenating old and straggling shrubs by mowing them, and conducting prescribed burns—will boost insect populations. Removing invasive shrubs and trees and replacing them with native species is hugely helpful because our insects tend not to feed on the leaves of invasive plants but thrive and multiply where they can eat the foliage of diverse native plants. Even on a small property—an acre or less—you can improve the habitat for flycatchers simply by replacing invasives with native vegetation.

WOODPECKERS

These birds can be easy to find and watch since their main mode of feeding—using their sturdy bills to chip away the wood of trees and expose insects—produces a steady, detectable sound. Males also drum to proclaim territories by rapidly striking their bills against limbs, hollow trunks, drainpipes, and building siding, causing a loud and resonant racket.

Woodpeckers chisel holes in living trees but rarely in healthy ones. They strip the bark off dead and dying trees to get at carpenter ants and the grubs of wood-boring insects, which can prevent those pests from spreading to nearby healthy trees. Woodpeckers also consume nuts, fruits, and seeds; tree sap is a major food for the yellow-bellied sapsucker. Several species, including the red-headed woodpecker, occasionally eat other birds' eggs and nestlings.

Woodpeckers nest in holes in trees, which they generally excavate themselves. Their holes also

Woodpeckers chisel holes in trees to get at ants and the larvae of wood-boring insects. They also excavate nest holes, cavities that are also used by other birds, mammals, and reptiles.
© *iStock.com/JP1961*

provide nesting and resting sites for smaller owls, bluebirds, tree swallows, nuthatches, chickadees, squirrels, and even tree frogs and snakes.

Nine kinds of woodpeckers live in eastern North America. The red-headed woodpecker is found throughout the East except for northern New England. The red-bellied woodpecker is the common woodpecker of the South and lives as far north as Wisconsin, Michigan, and New York. Yellow-bellied sapsuckers breed in the north and migrate to the Southeast to spend the winter. The downy woodpecker is a common permanent resident in woodlands across North America, as is the similar-looking but larger hairy woodpecker. The red-cockaded woodpecker lives in the Southeast and is now rare, local, and classified as an endangered species; its habitat, mature pine forest, has dwindled. The black-backed woodpecker, a species of the far north, lives among firs, spruces, and other conifers. The northern flicker eats many ants and is seen on the ground more often than other woodpeckers; it breeds throughout North America, with individuals from the north shifting southward in winter. The pileated woodpecker is found throughout the East and is the largest of all North American woodpeckers. It hammers rectangular-shaped holes in trees to get at nests of carpenter ants, a major food item.

Leave dead trees and snags standing so that woodpeckers can forage for grubs and excavate cavities. You can create feeding sites by girdling unwanted trees, which will then be invaded by wood-boring insects and ants, ultimately attracting woodpeckers. When planting trees for wildlife, choose types that produce foods used by woodpeckers. Red-headed and red-bellied woodpeckers, for example, eat acorns, so oaks should attract those two species to your property or yard. Woodpeckers eat the fruits of black cherry, black gum, hackberry, American holly, Juneberry, and red cedar, and those of sumac, elderberry, dogwood, and other shrubs. Several kinds of woodpeckers consume the purple fruits of pokeweed, a large weed that often grows in areas where the soil has been disturbed. Woodpeckers also eat fruits of Virginia creeper and poison ivy (flickers are especially fond of the latter).

If you don't have any large trees, put up nesting boxes, and woodpeckers may nest in them or use them for winter shelter. Like many birds, woodpeckers visit small sources of open water to drink and bathe; they prefer birdbaths that are tucked in among trees and shrubs rather than out in the open. A simple birdbath, such as a terracotta or stone bowl (or even an inverted metal or plastic trashcan lid) can attract woodpeckers.

GRASSLAND BIRDS

Productive grassland habitats are much rarer in the East than woodlands. Our region was heavily forested in the past, and it "wants" to be that way today, thanks to abundant tree species and seed sources plus good soil and climate conditions for growing trees. However, grasslands have also been a part of the environment for many centuries, on sites with droughty or poor soils and in the wake of fires.

Plants in grassland habitats can be short (six inches or lower), tall (four feet and taller), or a combination of both. The height of grasses and associated wildflowers, as well as their density, helps determine which grassland birds will use a given habitat, as does the presence or absence of scattered shrubs or trees. Native warm-season grasses, often called "bunch grasses," grow in clumps, letting birds walk through an area on the ground to find food and shelter. Imported cool-season grasses (the types growing in most hayfields and pastures and on sites such as capped landfills and airports) are somewhat less welcoming but still offer important habitat.

Many birds visit grasslands for food, including seeds and plant leaves as well as insects and other prey. Around 15 kinds of birds breed in eastern grasslands. The populations of those species have fallen over the last century due to changes in farming practices, land development, fire suppression, and forests taking over former grasslands. Most grassland birds nest on the ground. They're also "area sensitive," which means they won't nest in fields unless they're a certain size or larger. Bobolinks need fields that are at least 10 acres; eastern meadowlarks require around 20 acres; and Savannah sparrows need 20 acres. Grasshopper sparrows and northern harriers need 30 or more acres, and upland sandpipers prefer 150-acre sites with short, sparse grasses.

Other grassland birds include vesper, Henslow's, and field sparrows; dickcissel; bobwhite quail; ring-necked pheasant; eastern harrier; barn owl; and short-eared owl. Some grassland birds breed in northern areas and then migrate south for the winter, so grasslands are crucial habitats in both regions.

Farmers can help grassland birds by waiting to mow hayfields until after the birds' breeding seasons have ended, from the end of July to mid-August. (Some state wildlife agencies, the USDA Natural Resources Conservation Service, and certain wildlife organizations offer funds to help pay for delayed mowing. The Bobolink Project in the Northeast is one example; see the Wildlife Sketch "Bringing Bobolinks Back to New England" in Chapter 5.) If you have a sizable field and don't need income from haying, wait until September or even later to mow, and consider mowing every other year rather than annually.

Landowners can plant grasslands specifically for birds. Native warm-season grasses are somewhat harder to establish than cool-season grasses, but over time, they will need less fertilizer, lime, and mowing maintenance. They offer grassland birds better nesting cover and more dependable food sources, and they provide better winter cover since snow doesn't mat them down. Wildlife agencies and the USDA Natural Resources Conservation Service can provide advice and may share costs for planting warm-season grasses.

In grasslands, the height and density of grasses and other plants help determine the kinds of wildlife that will use the habitat. This field has a mix of cool-season grasses and herbaceous plants.

HUMMINGBIRDS

Although more than a dozen kinds of hummingbirds live in the West, only the ruby-throated hummingbird occurs east of the Mississippi. It winters in southern Florida, coastal Texas, and southward in Central America. In spring, hummingbirds usually arrive in their breeding areas, as far north as southern Canada, when early flowers start to bloom. In summer, a hummingbird will consume its own weight (three grams, or a tenth of an ounce) in nectar each day, along with many insects. Ruby-throated hummingbirds live in hardwood and mixed hardwood-and-coniferous forests, woodland clearings and edges, parks, and suburban gardens. Often found near water, they home in on areas where they can find nectar-producing flowers along with trees and shrubs where they can take shelter, perch, and build nests.

Hummingbirds insert their bills into flowers to feed on nectar; in the process, they pollinate many plants. They are especially attracted to bright-red blossoms, and some plants, including the woodland vine called the trumpet creeper,

A female ruby-throated hummingbird takes nectar from a bee balm flower. Through their feeding, hummingbirds pollinate many plants.
Tom Berriman

likely evolved red tubular flowers to attract hummingbirds. Ruby-throated hummingbirds take nectar from more than 30 types of flowers, including wild bergamot, bee balm, jewelweed, honeysuckle, Turk's-cap lily, and cardinal flower. Hummingbirds eat mosquitoes, gnats, small flies, and bees; they also pluck spiders out of their webs and glean aphids, small caterpillars, and insect eggs from the leaves and bark of trees.

Gardeners and folks establishing habitat plantings can select flowering trees, shrubs, vines, annuals, and perennials that hummingbirds will visit for nectar. Good choices include the following native varieties: cardinal flower, round-leafed and Carolina pink, columbine, phlox, skullcap, beardtongue, Oswego tea, Carolina silverbell, foxglove, trumpet creeper, jewelweed, pepperbush, rose mallow, azaleas, red buckeye, and rhododendron. Hummingbirds also visit the flowers of crabapples, hawthorns, redbuds, catalpa, and tulip tree. Consult field guides or your state university cooperative extension service to learn which of these plants are native to your area. Don't plant any that are not, as some, such as trumpet creeper, have the potential to become invasive in certain settings.

Hang up a hummingbird feeder in your yard or garden to attract these colorful birds. Both summer residents and migrating hummingbirds will feed on a sugar solution made of four parts of water to one part of sugar, which equals the average amount of sucrose in the nectar of flowers that attract hummingbirds. Boil the water and sugar for about two minutes (to slow down fermentation), and then refrigerate it before filling the feeder. Leave the feeder up as long as hummingbirds keep showing up; it won't delay their migration, and it will help late migrants stock up on energy so they can make it safely south.

12

Helping Other Wildlife

BUTTERFLIES

Among the most beautiful creatures in nature, butterflies pollinate many plants, and their caterpillar larvae are a key food for wildlife, especially birds. Several hundred species of butterflies live in eastern North America. Unfortunately, the populations of many have fallen in recent years. Humans have used pesticides and herbicides broadly and often carelessly, with pesticides killing butterflies and caterpillars directly and herbicides destroying plants they feed on. Development and intensive agriculture have erased butterfly habitat, and invasive plants outcompete many of the native plants on whose foliage caterpillars feed. As awareness of this situation expands, more and more property owners are planting butterfly gardens and wildflower fields to help butterflies and other pollinating insects.

Butterflies have four life stages: egg, caterpillar, pupa, and adult. Adult butterflies lay eggs on specific host plants. The eggs hatch into tiny caterpillars that feed on the plants' leaves. Cater-

Each autumn, eastern monarch butterflies make strenuous migrations all the way to Florida and even Mexico to spend the winter. You can create butterfly gardens and wildflower meadows where monarchs can feed on nectar from native flowers.
© iStock.com/blackwaterimages

pillars process a lot of greenery; most molt several times, shedding their exoskeleton (a support structure like a tough skin on the outside of the body) and growing a new one to accommodate their increased size. (Each caterpillar growth stage is known as an instar.) Next, the caterpillar goes into a dormant state, surrounded by a protective pupa, or chrysalis, inside which it metamorphoses over the next 10 to 15 days, changing into a butterfly. After adult butterflies emerge, they feed on nectar, the sweet liquid produced by flowers, and some also get sustenance and trace minerals from tree sap, rotting fruit, dead flesh, and animal dung. Males and females mate, female butterflies lay eggs on host plants, and the cycle repeats. One to three generations—also known as broods or flights—are typical for our eastern species each year.

Many adult butterflies live from one week to a month, with smaller butterflies having shorter lives. Some, including mourning cloaks and tortoiseshells, overwinter as adults, hibernating in cracks in wood, tree cavities, and behind bark; in spring, they emerge and breed. Monarchs famously migrate to warmer climes, such as Mexico. Other butterflies pass the winter in the egg stage.

Consult a butterfly field guide or check with your university cooperative extension service (or that of a neighboring state) to learn which butterflies live in your area and the plants they feed on. The leaves of oak and willow trees offer important food to the larvae of many species. Caterpillars of the spicebush swallowtail, a handsome black, blue, and yellow butterfly found throughout the East, feed on leaves of spicebush, sweetbay magnolia, sassafras, and prickly ash. Tiger swallowtail larvae eat the foliage of hornbeam, pawpaw, sassafras, spicebush, tulip tree, cherry, and wild plum. Monarch caterpillars feed exclusively on milkweeds. The caterpillars of the regal fritillary eat milkweed and thistle, while those of the great spangled fritillary eat violets, mainly at night. Mourning cloak caterpillars consume elm, willow, and poplar leaves. The red admiral eats nettles, and its close relative, the painted lady, feeds on burdock and everlasting.

While a caterpillar often feeds on the same plant and may even remain on a single leaf until it pupates, butterflies range more widely in search of food. They may take nectar from the flowers of many different plants. Butterfly mouthparts consist of a long tube, called a proboscis, kept coiled below the head when not being used to sip nectar. Nectar is sugar water with amino acids, vitamins, proteins, flavonoids, and enzymes. Many butterflies prefer red, purple, yellow, orange, or pink blossoms, especially those arranged in flat-topped or clustered flower heads. Butterflies are short-sighted and are attracted to large patches of color, so gardeners should plant solid masses of a single kind of flowering plant rather than a scattering of different flowers.

Good butterfly plants include purple coneflower, black-eyed Susan, woodland sunflower, bergamot, ironweed, butterfly weed (a type of milkweed), other milkweeds (many species exist in the East), asters, blazing stars, joe-pye weed, rose verbena, wild geranium, and cardinal flower. Most of these plants need at least six hours of full sunlight per day. For a butterfly garden, use caterpillar plants as the core and add flowers to feed adults. Pick several different kinds of flowers that don't bloom all at once but rather in sequence so that butterflies can always find nectar. Native plants are far and away the best choice for butterfly gardens because butterfly caterpillars have adapted to feeding on them over thousands of years. Some butterflies have coevolved with specific plants and pollinate their flowers. Native plants tend to be hardy and can tolerate drought, excessive moisture, cold, and heat. Purchase pure native stock rather than cultivars that have been bred for bigger or showier flowers, as the latter may have lost the fragrance, nectar, or structural features that attract butterflies. You can also collect seeds from native flowering plants growing in the wild locally.

Despite its attractive name, one flowering plant to avoid is the so-called butterfly bush (*Buddleia* species, from Asia). Widely sold by garden supply companies, butterfly bush has proven to be an invasive. Adult butterflies take nectar from its showy flowers, but their caterpillars won't feed on the plant's foliage. Butterfly bush has no natural insect predators in North America, and the plants spread far and wide via tiny, wind-borne seeds.

Like all insects, butterflies are cold-blooded. Their bodies and muscles need to warm up enough so they can become active. Most species start flying when the air temperature is about 60 degrees Fahrenheit. Butterflies also have a hard time flying in wind. Site butterfly gardens in sunny areas sheltered from the wind. A screen of shrubs will blunt the wind and also put forth nectar-producing blossoms. Pick shrubs that are native to your region. Good candidates are spicebush, fragrant sumac, meadowsweet, New Jersey tea, nannyberry, black haw, mapleleaf viburnum, buttonbush, hawthorn, oakleaf hydrangea, Virginia sweetspire, and beautyberry.

Place flat stones on the ground to soak up heat from the sun; butterflies will bask on the stones to elevate their body temperature, allowing flight. Butterflies need shade when it's hot. Males of some species gather in wet spots, including moist sand and the edges of puddles, where they pick up minerals and other nutrients that they pass on to females during mating. Install a water dripper over bare sand, soil, or gravel, then add salt (table salt is adequate, but sea salt has more trace elements), wood ash, manure, rotting fruit, or stale beer. Experiment with different substances to see which ones attract butterflies in your area.

Avoid using insecticides in or near a butterfly garden. Chemicals such as malathion, carbaryl (brand name Sevin), and diazinon kill butterfly caterpillars. Bt is a preparation that includes a bacterium, *Bacillus thuringiensis*, and is often marketed as an "organic" insecticide that does not harm beneficial insects such as honeybees and ladybugs. However, it's deadly to butterfly caterpillars.

Plant a butterfly garden, and you'll have less lawn to mow. You will also be treated to beautiful colors from flowering plants and the different kinds of butterflies that visit their blooms.
Howard Nuernberger, Penn State

Many books provide details on how to identify, attract, and enjoy butterflies. They include

Butterfly Gardens, by Alcinda Lewis (Brooklyn Botanic Garden, 2007); *Gardening for Butterflies*, by the Xerces Society (Timber Press, 2016); *Attracting Native Pollinators*, by the Xerces Society (Storey Publishing, 2011); and the field guide *Butterflies Through Binoculars: The East*, by Jeffrey Glassberg (Oxford University Press, 1999), which also offers advice on observing and photographing butterflies. The Princeton Field Guide series includes *Caterpillars of Eastern North America: A Guide to Identification and Natural History*, by David Wagner. The websites of the National Wildlife Federation, the North American Butterfly Association, and the Xerces Society for Invertebrate Conservation have excellent information.

POLLINATING INSECTS

In addition to butterflies, other pollinating insects need more acres of healthy habitat, including native bees, wasps, flies, moths, beetles, and bugs. As they feed on nectar, many of these insects transfer pollen grains from flower to flower and plant to plant, fertilizing flowers so that they develop into fruits and nuts, food for a range of wildlife. Insects, especially honeybees and native bee species, also pollinate many domesticated crops—a natural "service" valued in the billions of dollars. Farm products pollinated by insects include apples, pears, nuts, strawberries, tomatoes, peppers, blueberries, squash, and melons.

In recent years, populations of domesticated honeybees have declined sharply across North America, with entomologists attributing cases of colony collapse disorder, or CCD, to diseases, parasitic mites, malnutrition, and neonicotinoid pesticides. Because of this loss, wild pollinating insects have become increasingly important to our farm economy. Unfortunately, populations of wild pollinators also have fallen, likely from pollution, diseases, stress brought on by climate change, improper use of herbicides and pesticides, and a loss of food plants caused by development, intensive farming, and invasive plants taking over many habitats.

Nearly 4,000 species of wild bees live in the United States. Some live in colonies, while others lead solitary lives. Among the better-known types in the East are bumblebees (*Bombus* species), which visit wildflowers and crop and garden plants. Bumblebees are colonial insects that mainly nest underground. Other important pollinators include sweat bees, carpenter bees, digger bees, mason bees, and leafcutter bees. Moths visit flowers for nectar mainly late in the day and at night. Many flies also pollinate wildflowers and the flowers of shrubs and trees.

Mowed expanses of lawns attract few pollinating insects; gardens and fields of wildflowers offer more and better resources. To create habitat for pollinators, pick a site that gets plenty of sunlight, such as part of your backyard or an old field that you don't want to keep mowing. The soil over a septic drain field, strips of grass around parking areas or athletic fields, roadsides, and water discharge areas are all good places in which to establish plots of nectar-producing plants. First eliminate existing cool-season grasses, either by solarization (laying down fabric, plastic, or other coverings that kill grass plants by overheating them and depriving them of moisture), mechanical means (sod cutters can be rented in most areas), or applying herbicides. Dig or rake out and dispose of grass roots, leaving a smooth, weed-free soil surface in which wildflower seeds can be bedded. Buy seeds from growers who specialize in "local eco-type" stock—harvested, collected, or produced locally from native varieties and adapted to your area's soils, insects, plant diseases, and climate.

The Pollinator Partnership is an international nonprofit organization that works to promote the health of pollinators through conservation, education, and research. A series of science-based regional planting guides, *Selecting Plants for Pollinators*, can be downloaded from the Partnership's website. They also offer guides covering specific ecoregions in the East, such as the Adirondacks and New England, the Central Appalachians, the Atlantic Coastal Plain, the Ozark Mountains, the Southeast, the Upper Midwest, and so on.

Another nonprofit organization, the Xerces Society for Invertebrate Conservation, has produced a series called Pollinator Habitat Installation Guides, including *Establishing Pollinator Meadows from Seed, Organic Site Preparation for Wildflower Establishment*, and *Collecting and Using Your Own Wildflower Seeds*. These detailed references can be used in combination with region- and state-specific guides, also from the Xerces Society, to plan, create, and maintain nectar- and pollen-rich habitats such as wildflower meadows, shrub hedgerow plantings, and conservation cover. Download these resources from the Xerces Society website. The USDA Natural Resources Conservation Service (NRCS) can help you plan plantings for pollinators, and they offer cost-sharing for some projects.

A pollinator field is an excellent project for a school, church, or garden club. Farmers, town and municipal conservation commissions, and landowners can also plant for pollinators. Businesses are getting into the act as property managers realize that pollinator plantings are easier and cheaper to maintain—and much more interesting to look at—than manicured lawns.

DRAGONFLIES

Brightly colored dragonflies hunt for other insects above ponds, lakes, streams, rivers, wetlands, and sunny glades and fields. They are closely related to damselflies, which are smaller than and not as stocky as dragonflies and are neither as fast nor as agile in flight. A dragonfly starts its life when it hatches from an egg laid in an aquatic habitat. The resulting drab-colored larval nymph lives in and among underwater plants. A typical eastern dragonfly lives for about a year, spending approximately 11 months as a nymph before metamorphosing into a flight-capable adult that leaves its watery habitat and lives for roughly another month. In both nymph and adult stages, dragonflies are carnivores. Nymphs eat aquatic larvae of mosquitoes, stoneflies, caddis flies, mayflies, and other dragonflies, plus water beetles, snails, and small fish. Adult dragonflies catch other insects in flight, including blackflies, deerflies, horseflies, midges, mosquitoes, and swarming ants. A single dragonfly can eat hundreds of mosquitoes a day. In turn, dragonflies are preyed on by fish, frogs, and birds.

Dragonflies need unpolluted waters where their nymphs can develop. Diverse native plants and shrubs edging such waters will provide food and cover for dragonfly prey insects. Dragonflies will visit backyard ponds whose surfaces have some open water. (For information on building small ponds, see Chapter 10, "Habitat Around Your House.") Landowners can also create larger seasonal and permanent wetlands and ponds that dragonflies will quickly colonize.

SALAMANDERS

Salamanders are amphibians with moist, smooth skin. Most are small, secretive, and rarely seen. Their life stages are egg, larva, juvenile, and adult. Most eastern salamanders have aquatic eggs and larvae, followed by a land stage for the juveniles and adults. (Not all species follow that sequence: Some salamanders spend their whole lives on land, while others dwell only in water.) The Southern Appalachian region in particular hosts a tremendous number and variety of woodland salamanders, including some species that have very small ranges and are found nowhere else in North America or the world.

During all growth stages, salamanders are carnivorous, with the size of the prey determined by the size of the salamander. Salamanders eat insects and other invertebrates. On land, they find food in the forest's leaf litter, beneath the soil, under rocks, in rotting wood, and in shrubs and trees.

In the Words of Wisconsin Landowners

AMBER ROTH, A WILDLIFE MANAGEMENT PROFESSOR at the University of Maine, studies the breeding ecology of the golden-winged warbler. That small, colorful songbird nests in the Appalachians and the Upper Midwest, in young forests growing back following fires, windstorms, and clear-cut logging.

Roth belongs to the Golden-Winged Warbler Working Group, a coalition of biologists and habitat managers in North and South America seeking to boost this bird's population, which has fallen drastically over the last several decades. Roth helped organize the Wisconsin Young Forest Partnership, or WYFP, in which more than a dozen partners—including federal and state agencies, wildlife organizations, and timber companies—are reaching out to landowners to explain how animals such as golden-winged warblers need forests younger than the middle-aged woods that now cover many parts of the Great Lakes states. A dearth of young forest is causing populations of many kinds of wildlife to decrease by a small percentage every year. Those animals may not be endangered yet, but they've become rare enough that many states, Wisconsin included, consider them "species of greatest conservation need."

Says Roth of the Wisconsin Young Forest Partnership: "One objective is to educate and engage landowners who are not currently managing their forests so that they'll consider all of their options and, we hope, decide to make and enhance young forest for high-conservation-priority wildlife like golden-winged warblers."

Another objective is to promote a healthy timber industry that will provide jobs in rural Wisconsin. Losing that industry would hurt local economies. It would also remove a cost-effective tool for creating young forest habitat, Roth says.

WYFP offers professional site visits and habitat management plans, then helps landowners apply for funding to get projects completed—timber harvests, and also activities such as shearing alders, in which a tracked machine chews down old, over-mature alder shrubs, stimulating them to grow back more thickly, offering better feeding and nesting habitat for golden-wings, American woodcock, snowshoe hares, and other young forest wildlife.

Says Roth, "Landowners need people they can talk to, people they can ask 'Who did you hire to get the management done?' and 'How did it turn out in the end?' Participating landowners can network with other landowners and connect with consulting foresters, biologists, and other conservation specialists who can give good advice on how to manage the land for young forest."

Many Wisconsin landowners like to hunt. Others simply enjoy seeing diverse wildlife on their properties. The following landowners explain why they've joined the Wisconsin Young Forest Partnership.

Les Strunk, from Oconomowoc: "What went through my mind was this: If I make the habitat better for birds, it'll also be better for deer. The alder shearing gave me

Veronica Berg (right) visits with equipment operator Mike Riggle, who is taking a break from mulching older shrubs and trees to spur the regrowth of thick new habitat on the Berg property in northern Wisconsin.
Todd Berg

an opportunity to go in and open up old trails, plus put in new trails to get around on my property."

Peter Ourada, from Antigo: "We plan to continue using timber harvests to make our property a productive woodland where we can conduct other commercial harvests in the future. What we're doing will help wildlife, and it will help us, too, by creating better hunting conditions along with more opportunities to view wildlife. What I've learned has helped me become a better landowner and get more enjoyment out of my land."

Mike Gardner, from Rusk County: "The work gave me a management toehold. After that initial push, I can now do a lot of the continuing habitat work myself, including mowing with a tractor to keep the shrubs young and vigorous. The project has let me renew my personal commitment to having a diversity of wildlife habitats of different ages on my land."

The golden-winged warbler, American woodcock, and New England cottontail are all emblematic of young forest. Create habitat for any one of that trio and—depending on where you live—you'll have a good chance of helping up to 60 other kinds of reptiles, birds, and mammals, including box turtles, wood turtles, green snakes, ruffed grouse, rusty blackbirds, American redstarts, snowshoe hares, and Canada lynx. Pollinating insects, monarch butterflies, and dragonflies use those habitats, too. And bats swoop over young forest stands to catch prey.

Dan Eklund is a wildlife biologist with the U.S. Forest Service on the Chequamegon-Nicolet National Forest in northern Wisconsin. "To have a healthy ecosystem, you need to have some young forest around," he says. "A lot of animals that breed in mature forest also need young forest at one time or another—and a lot of animals that breed in young forest need mature forest as well. The key is keeping all the different elements in the right balance."

Some salamanders breed on land; others lay their eggs in the waters of temporary vernal pools. Salamander populations peak in older forest where trees produce deep shade, and fallen trees and limbs and rotting logs lie on the ground.

Many salamanders breed in vernal pools that fill with water from rain and snowmelt in spring. If you have a vernal pool on your land, take care of it. Do not try to dig it deeper because that will ruin it. Destroying a vernal pool would likely cause the amount of biomass, or animal life, on your land to plummet. You can create a vernal pool and improve your land as a salamander habitat.

Salamander populations are often high in areas of older mature forest, especially where the tree canopy produces deep shade and where fallen trees and rotting logs litter the ground. Clearcut timber harvests can set back local salamander populations. In studies conducted in North Carolina, Missouri, and Pennsylvania, wildlife researchers found that numbers of forest-dwelling salamanders declined following clearcutting. When all the trees in a forest stand are removed, direct sunlight heats the ground and drives down moisture levels in the leaf litter and soil, creating conditions that many salamanders cannot tolerate. It can take at least seven and sometimes 20 or more years for salamander populations to rebound following a clearcut. If you manage a forest for timber and wish to preserve salamanders, use selective timber harvests that keep significant numbers of trees on a site to cast some shade. Leave a few large logs on the ground, do not cut heavily around spring seeps, and leave buffers of standing trees around streams and vernal pools.

Acid rain (also called acid deposition, since snow can be acidic as well) threatens salamanders and other amphibians. Many vernal pools are in forests where soils are already acidic because of underlying rock types. When acid rain further lowers the pH of vernal pools, larval salamanders may not develop into healthy juveniles that can grow quickly enough to leave the pool before it dries up. Conservation-minded citizens need to be politically aware and resist any attempts to weaken the Clean Air Act of 1970 and other anti-pollution regulations that have slowly resulted in less-acidic precipitation. It's critically important to wildlife and forest health that we keep our air and water as pure as possible.

TURTLES

Eastern turtles show an impressive ability to use many different habitats: streams, rivers, swamps, bogs, grasslands, dry sandy areas, shrublands, mature forests—one or more species of turtle can be found in all of those settings. Many turtles spend much of their lives in freshwater. In moving water, most stick to shallows where the current isn't strong. Water-dwelling turtles need plants to eat—plants that attract and nourish prey, and plants that provide cover where turtles can hide to ambush prey or, when they're small and vulnerable, to avoid their own predators.

Cattails are native plants that, if they grow too thickly, can make aquatic habitats too dense for turtles. Thin them by hand-pulling, mowing, and cutting (if you cut cattails below the water line two or three times in a season, fewer will grow back the following year). Cattails can also be suppressed by drawing down a pond's water level in autumn so that freezing damages the plants' roots over winter. Purple loosestrife is a non-native invasive that chokes aquatic habitats; digging out the plants or using biological controls or herbicides can check infestations.

Aquatic and semiaquatic turtles need places where they can haul themselves out of the water and bask in the warming rays of the sun. Fell streamside trees so that their tops and upper trunks lie in the water of a beaver pond, river oxbow, or slow-moving stream, and their lower trunks project out of the water. Planting a buffer zone of native shrubs and low plants along a stream or around a pond or wetland will slow and purify surface water before it enters those water features, protect their earthen banks from erosion, and improve food and cover resources for turtles.

Land-dwelling turtles are in trouble as their living spaces are cut up by roads and development. Many turtles that people see today, particularly in highly developed parts of the East, are members of what ecologists call "living-dead" populations, composed of older adults that have a hard time reproducing successfully due to high mortality rates in their eggs and hatchlings. Landowners can create and preserve as much habitat as possible and make sure corridors exist so that turtles can move safely from one part of their habitat to another. A book that I wrote, *Turtles: Wild Guide* (Stackpole, 2007), has a chapter "How You Can Help" that offers detailed suggestions.

Eastern box turtles live from southern New England to Florida. Although mainly land-dwelling, box turtles will swim across ponds and streams when traveling and will soak in water during hot weather. They feed on insects, snails, slugs, earthworms, spiders, and other small invertebrates; plant foods include roots, shoots, buds, leaves, and fruit, especially wild berries. A mosaic of different habitats fairly close to one another will favor box turtles: streams, ponds, vernal pools, shrublands, young forest, forest edges, glades, and older forest. When harvesting timber or cutting back shrubs to rejuvenate them, work in winter when box turtles are hibernating in soils, woody debris, and old mammal burrows.

Wood turtles are found in the Northeast. They spend the winter dug into sandy bottoms of smaller headwater streams, then they feed and reproduce in nearby riparian zones in spring, summer, and fall. Timber harvests can improve feeding opportunities for wood turtles by creating openings in the forest canopy that let in sunlight, spurring the dense growth of forbs, shrubs, and young trees, producing more abundant fruit and a greater prey base of insects. Protecting wetlands by planting generous upland buffers around them will help spotted turtles, which forage for food mainly in water but also venture onto land. Females lay their eggs in loose sandy soil in sun-bathed open areas, including old fields and forest clearings.

BATS

Bats are our only mammals that fly. (Flying squirrels actually glide.) Bats feed voraciously on insects, taking their prey in flight. A single bat may scoop up 500 insects in an hour and close to 3,000 insects in one night, amounting to a quarter of its body weight. Bats eat moths, leafhoppers, flies, wasps, mosquitoes, midges, bugs, and beetles, including many agricultural pests. They often feed over open water, where they can also drink.

Bats need different types of habitat depending on the species of bat, the season of the year, and their activities: feeding, raising young, resting, or hibernating. Promoting native vegetation will boost the number of insects that bats feed on. Creating clearings in wooded areas, making a seasonal or permanent pool, and planting buffers of vegetation around water will help bats, as will maintaining old fields and shrublands and preserving trees with cavities.

Bats sleep and rest in caves, mines, crevices in rock outcrops or cliff faces, tree cavities, humans'

Bat houses should face south or east and receive six to eight hours of direct sunlight daily, which helps keep their inside temperatures warm. The best locations are near streams, rivers, lakes, and forest edges, where bats find plentiful insect food.
© *iStock.com/erniedecker*

buildings, and behind loose bark on trees. In early summer, pregnant females gather and give birth to young, typically in warm, secure places such as tree cavities, house attics, and bat boxes. Often, groups of bats return to the same maternity site every year. In winter, most bats hibernate in the cool, stable temperatures of caves, abandoned mines, cemetery crypts, and rock crevices.

Putting up a bat roosting box will help local bats. Most boxes are one to two feet tall and have layers of baffles inside. Boxes 12 inches in height can accommodate up to 100 bats, and boxes 24 inches in height can house as many as 200 bats. Bat Conservation International (BCI) offers design plans for three different-sized bat houses on its website, plus tips on how and where to install a bat house: ideally, on an east- or south-facing side of a building that gets six to eight hours of direct sunlight daily. You can also put up a bat house on a pole. Site houses 20 to 30 feet from tree branches or other obstacles and 12 to 20 feet above the ground. The best locations are along streams, rivers, and lakes, and near forest edges. BCI lists vendors that sell prebuilt bat houses certified to meet bats' needs. More information is available from *The Bat House Builder's Handbook* by Merlin Tuttle, Mark Kiser, and Selena Kiser (University of Texas Press, 2005).

If you have bats in your area, they may not move into your bat house right away because they're already using other roosts. In areas where roosts are scarce, you may quickly attract bats. Surveys of bat box owners suggest that the overall occupancy rate is just over 50 percent, with larger boxes used more often than small ones.

Since 2006, a fungal disease called white-nose syndrome has killed millions of bats and severely cut into local populations in many parts of the East. The fungus grows on hibernating bats, causing them to waken and burn calories, lose body fat, and starve. By creating high-quality feeding and breeding habitats for bats, landowners can give these important and intriguing mammals their best chance of surviving this threat.

RABBITS AND HARES

Five cottontail rabbit species live in eastern North America. By far the most common is the eastern cottontail, found region-wide except for northern New England and considered an introduced species in many areas. The swamp rabbit inhabits parts of the Deep South; the Appalachian cottontail lives in scattered sites in the Appalachian Mountains; the marsh rabbit is found on the coastal plain from Virginia south to Florida; and the New England cottontail lives in portions of five New England states and in eastern New York. The snowshoe hare is a larger member of the same family, Leporidae, and it ranges from Canada and Maine to North Carolina in the Appalachians and also lives in the northern Great Lakes states.

All of these creatures mature quickly, breed several times a year, and have the potential to produce many young. They represent an important food source for hawks, owls, fishers, foxes, coyotes, bobcats, and Canada lynx. Swamp and marsh rabbits have a strong affinity for wetlands and are able swimmers; New England cottontails also use wetland habitats, as do snowshoe hares. Appalachian and New England cottontails inhabit scrub oak and mountain laurel thickets and young forest. As prey animals and vegetarian feeders, rabbits and hares need dense cover in which to hide combined with abundant plants to eat. Summer favorites are grasses, low broadleaf plants, and clovers. In winter, cottontails feed on woody shoots and bark.

Excellent cottontail habitats include old fields, shrublands, woods borders, and forest stands that have been logged within the past 10 to 15 years and where young trees grow back thickly along with native shrubs and vines. Swamp rabbits feed heavily on cane, and marsh rabbits eat aquatic emergent plants including cane, cattails, and rushes. Cottontails take shelter in thickets and

WILDLIFE SKETCH

Respect

When deer are in their autumn rut and bucks focus on mating with does, they fight with one another to establish dominance, lose their instinctive caution, and, very rarely, act aggressively toward humans.
© *iStock.com/FRANKHILDEBRAND*

WITH A .22 RIFLE IN MY HANDS, I CREPT THROUGH THE WOODS. I looked about for acorns and hickory nuts freshly chiseled by incisor teeth, for the flick of a bushy tail behind a tree's trunk. I listened for rustling in the leaves and the hollow knocking calls of gray squirrels in the treetops.

I saw a sapling's branches waving back and forth and heard sounds of grunting and stamping.

I spotted the gray-coated body of the buck at just about the same moment that he saw me. He had been beating on the sapling's truck with his antlers, which had three sharp points on each side. His neck was swollen from the rut. His eyes bulged, and his nostrils flared. I noticed the hair stand up on the back of his neck and felt the hair stand up on mine.

The buck grunted again. He lowered his head and took a couple of steps toward me. I began backing up, keeping my eyes on the deer. My heart pounded in my chest, and my vision blurred at the edges.

A log lay on the ground between us. It was maybe 20 feet long and about a foot thick, positioned on a diagonal between the deer and me.

The buck raised his head and stared at me. Then he ran forward. I shouted, and what came out of my mouth was something between a cry of outrage and a scream of fear.

The buck chose to pass on the far side of the log. Three long paces separated us. I got a brief whiff of his rank scent as he trotted past.

As the buck ran off along the mountain bench, I went and sat on the log. My heart was pounding. I believed that the log had diverted a potential charge by the buck. Or maybe my shout had persuaded him that I was a human and not a rival buck. There was no way of knowing for sure. I've read about hunters being charged, knocked down, and even gored by rut-crazed bucks who, at any other time of the year, would have run fearfully in the opposite direction. The hunters had done nothing to provoke the attacks.

I considered the strength of a 150-pound man versus that of a 150-pound buck, and I knew it would have been an uneven contest. The .22 rifle might have made a difference—and maybe not.

It pays to have respect for wildlife. I don't simply mean respect for a creature's ability to blend in with its surroundings, or find food in winter, or detect scent, or see in the dark. I mean respect for its raw power, its animal strength, and its wildness.

Patch cuts of two to five acres in woodlands will increase the amount of browse and cover available to rabbits and hares. In this small clear-cut, young aspen trees are growing in thickly. Songbirds, ruffed grouse, bobcats, moose, and deer also use patch cuts.

among young, low-growing conifers; snowshoe hares use evergreen areas frequently.

Patch cuts of two to five acres in woodlands will increase the amount of browse and cover available to rabbits and hares. Songbirds, ruffed grouse, bobcats, moose, and deer also use such cuts. Landowners can use income from selling timber for other habitat enhancements, such as planting native shrubs.

General approaches to encouraging cottontails are to renew shrubland habitat by mowing or cutting shrubs when they get too leggy and thin so that they grow back more thickly; building brush piles in different habitats and along travel corridors; planting patches of cool-season and warm-season grasses; mowing strips in grassy areas to stimulate succulent low growth; daylighting apple trees in old fields and orchards; maintaining old-field habitats; and logging woods near old fields and shrublands. Create a feathered edge by cutting trees where a forest meets a grass or weed field. The width of the cutting (from the field back into the woodland) can vary from 25 to 50 or more feet. Leave the tops and limbs of cut trees on the ground or build them into brush piles. To stimulate new sprouting, recut borders every five to 10 years.

Cottontails will use invasive shrubs, particularly multiflora rose, as hideaways to avoid predators. Before launching an all-out effort to suppress invasives in a shrubland, plan the work in several stages, always leaving enough thick shrubs for local cottontails to use for escape cover. In the same vein, if you mow or cut native shrubs to stimulate their regrowth, never treat an entire habitat area at the same time; instead, leave enough shrubs uncut so that rabbits and hares have some thick structure in which to hide.

WHITE-TAILED DEER

Many landowners hunt deer and want to improve their properties to attract and hold these game animals. Deer are also beautiful and interesting to observe in all seasons. Like all wildlife, they need dependable sources of water, food, and cover. Deer get much of the water they need by feeding on plants, and they also visit and drink from open water, including constructed ponds. Deer graze on many different kinds of grasses and herbaceous plants, browse on the buds and twigs of shrubs and small trees, and feed heavily on high-energy mast, especially acorns. Deer can be active at all times of the day, but mostly they rest in thick cover during daylight hours and feed in more open habitats at dusk, periodically through the night, and at dawn. They bed down in cool areas when it's hot, in sunny areas during cool weather, and in protected sites when it's cold and snowy.

You can create and improve food sources and sheltering opportunities by using a chainsaw to selectively cut vegetation. If you have some oaks, remove competing trees so that the oaks spread out their crowns and produce abundant acorns. Cut away trees that are shading out apple trees (deer love to eat windfall apples) in shrublands and old fields. Scatter patch cuts of varying sizes (two to five acres are optimal) and irregular shapes in wooded areas; you may be able to sell the trees to a logger or a firewood supplier. Tangles of briars, grasses, flowering plants, shrubs, and young trees will spring up in these new openings, providing food and secure spots where deer can bed down. Cuts sited on sunny south-facing slopes will attract deer in winter. After logging, trees such as aspens, red maples, and oaks will sprout vigorously from their stumps or roots, and deer will feed on these shoots.

In a pole-stage forest stand, trees aren't dense enough to offer good cover to deer, and they're not old enough to produce nuts. Hinge-cutting is a technique that can quickly turn such a stand into a better deer habitat. A foot or two above the ground, saw partway through the trunk, then push the tree over the rest of the way so that it remains partly attached at the stump and with its crown resting on the ground. Deer will browse the leaves and twigs and find good hiding cover among the slanting trunks. The trees will remain alive and keep sending up shoots. Deer particularly love places where they can bed down and feed in the same area.

Plant warm-season grasses for grazing; doe deer will hide their young fawns among the clumps of tall fountain-shaped grasses. Broadening hedgerows and fencerows, either by planting or allowing native shrubs to grow, will yield a thick habitat corridor that deer will readily use. A 50- to 100-foot-wide buffer of grasses, shrubs, and trees planted next to or on each side of a waterway will become a corridor and a feeding habitat. Planting patches of evergreens—cedars, pines, spruces, and firs—creates thermal cover where deer shelter during winter's snow and cold. Experts recommend plantations of five acres and larger, whose trees will provide excellent thermal cover when they reach 10 to 15 feet high. In areas where deer are already plentiful, you may need to fence conifer plantings so that whitetails don't browse them and prevent them from growing.

Blueberries are moderately shade-tolerant, but cutting a few trees to let in more sunlight can spur these woodland shrubs to spread and to produce more fruit. Black bears, ruffed grouse, songbirds, mice, chipmunks, and land turtles are a few of the animals that eat blueberries.
© *iStock.com/TT*

BLACK BEAR

Black bears are dedicated omnivores and serious landscape-travelers. Unless you own a huge parcel of land, you won't keep individual bears as fulltime residents. However, if you do own a fairly large tract that's mostly wooded, you probably enjoy the thrill of seeing a bear every now and then—and there are things you can do to increase those sightings. Black bears are found in the Upper Midwest, New England, the Mid-Atlantic, the Appalachians, and parts of the South. Ideal habitat is fairly inaccessible terrain that mixes wetlands, woodlands, shrublands, and some agricultural land. Bears frequent forest stands that include oaks, hickories, black cherries, and other trees that produce hard and soft mast.

Bears eat grasses, forbs, and the fresh foliage of shrubs and trees in spring. In summer and fall, they feast on Juneberries, blueberries, huckleberries, raspberries, blackberries, grapes, cherries, apples, acorns, hickory nuts, hazelnuts, and beechnuts. Bears in Florida eat the buds and fruits of saw palmetto and the berries and buds of cabbage palm. Bears eat carrion whenever they can find it; in spring, bears are serious predators of white-tailed deer fawns. Since bears are so highly omnivorous, follow a general strategy of creating diverse food sources based on native shrubs and trees. Selective timber harvesting can increase the percentage of mast-producing trees in a woodlot or forest. As with deer, you can create wooded or brushy travel corridors that link feeding areas.

Bears use wetlands extensively, feeding on the lush vegetation (highbush blueberries are a favorite) and hiding in these less-than-passable places during hunting season. In winter, bears hibernate in caves, dens dug beneath uprooted trees and under stumps, beneath brush piles, under logs, and in hollow logs. Protect large old trees that develop cavities and hollow trunks in which bears may den.

13

Enjoy Your Habitat!

SOME FOLKS LOVE THE PHYSICAL WORK they do when making homes for wildlife—whether it's running a chainsaw, walking along behind a brush mower, levering invasive honeysuckles out of the ground, or planting shrub seedlings or wildflowers. Enjoying what you've created is important, too. The best way to do that is to get out on the land. Just as you spent time carefully evaluating your property before deciding which habitat-improvement projects to undertake, and as you devoted hours to carrying out projects or overseeing others' efforts, so too should you spend time reaping the rewards of making new and more habitat for wild animals.

While carrying out habitat projects, consider making or expanding a network of travel arteries for you, your family, and your friends to use. You can think of these byways as woods roads, paths, and sneaks. You'll use these tracks for years to come as your habitat areas develop and as you

Skid roads put in during timber harvests make great paths for wildlife watching and grassy food strips that attract deer, turkeys, ruffed grouse, and other wildlife. This landowner exercises his German shorthaired pointers on lands managed for woodcock and grouse.

attract new and more wild creatures to your land.

A woods road should be wide enough to accommodate a pickup truck or similar vehicle (and, as such, great for harvesting and bringing in firewood and conducting future habitat work).

A path, to my way of thinking, is narrower and more apt to wind about, wander, climb steep grades, or thread through damp areas, visiting wildlife hotspots from beaver ponds to warm-grass plantings to field edges.

A sneak? It can be as basic as a lightly improved deer trail that permits quiet walking through woods, shrublands, or wetlands.

Making a woods road from scratch generally requires heavy equipment. You may be able to integrate creating roads with carrying out habitat projects, especially if you have a bulldozer or a log skidder on-site. Or you may already have such roads on your land, and they're simply overgrown or obstructed by blowdowns; in that case, a chainsaw and some pleasant hours out-of-doors can resurrect a network for work or recreation. Later, add paths and sneaks that branch off from or connect your main roads. Remember to add these arteries to your property map.

A great time to locate an existing woods path—such as a former skid trail used during a long-ago timber harvest—is after a light snowfall. Under such conditions, your eyes will be able to pick up faint courses that, dusted with snow, appear whiter and more level than the surrounding land. Use colored flagging to mark the path in its entirety. Then return with a chainsaw, a bow saw, or a belt saw, plus heavy-duty loppers and a mattock or Pulaski tool (a tool with a pick-axe blade on one end of its steel head and an axe bit on the other). Remove saplings and grub out shrubs such as mountain laurel, scrub oak, or blackberry. Toss aside fallen logs and branches, and lop off limbs that jut out into the path.

A sneak, the most rudimentary of byways, is also the easiest to make. You are basically improving on a game trail. Again, a dusting of snow makes such trails more visible. Deer, coyote, or fox droppings or tracks can also give away these natural wildlife routes. Mark the sneak with flagging, and then make minimal improvements with loppers, a belt saw, and maybe the Pulaski tool. Feel free to skirt the thickest tangles or leave a few fallen tree trunks to step over. As you gradually improve and extend a sneak, you'll be rewarded with an understanding of how animals pick the path of least resistance when negotiating slopes, shrub thickets, wetlands, and other habitats. Moreover, you will have a trace that, with usage, will become increasingly obvious and easy to follow.

Creating trails and sneaks is one of the most satisfying of outdoor endeavors since few other tasks yield so much progress for a minimal amount of time and effort. Even more fun than making and maintaining roads, paths, and sneaks is using them for hunting, hiking, snowshoeing, and watching wildlife.

PRODUCTS FROM YOUR HABITAT

Harvest natural products from habitats you create or enhance. Wild jams, jellies, and sauces can be made from apples, crabapples, wild grapes, cranberries, elderberries, blueberries, blackberries, black raspberries, Juneberries, wild cherries, sassafras, mints, sumac, and violets. Wild nuts such as butternuts, black walnuts, shagbark hickory, pecans, American chestnuts, and hazelnuts are delectable. Pick fern fiddleheads and ramps in spring and wild mushrooms in summer and fall. Foraging is a great way to learn about trees and shrubs, and you'll often encounter wildlife while collecting wild foods.

Many landowners harvest wood when creating a younger forest growth stage or making a feathered edge between forest and field. Financial returns from timber harvests can pay for more habitat improvements, such as buying and planting native shrubs. Some landowners sell firewood to bring in a modest but steady flow of income.

Heating your house with wood cut on your own land is very satisfying—something I can vouch for, having done it for more than 30 years.

EQUIPMENT AND GEAR

Some folks like to use a UTV—a utility task vehicle—to get around on their woods roads and paths. My wife says I'm allowed to get one when I'm an old man. (I'll be 67 this year, and I don't want to admit to being old, even though I would dearly love to get a John Deere 6x4 Gator—a very stable machine in which you can ascend and descend hills safely. It looks pretty, too, with that bright green paint and cheery yellow seat.)

However, shank's mare is really the way to go. Every now and then, you'll want to rest, either to have a sandwich, pay some extra attention to your immediate surroundings, or let things quiet down so you have a better chance of seeing wildlife. Consider buying a lightweight unipod or tripod seat that you can carry in a daypack or fanny pack, then quickly set up to take a load off your feet. Along game trails, near wildlife focal areas such as spring seeps or vernal ponds, or in places where you've created or enhanced habitat, set up for an hour and, comfortably ensconced on your portable chair, wait for wildlife to appear. Portable chairs also elevate you above ticks in leaves or low grass. (Always use caution if you live in areas with ticks—which, unfortunately, now amounts to almost all of the East.) You can fell trees in promising spots, giving yourself a convenient stump or log to sit on and adding some dead and downed wood to the habitat. Cutting trees can open up views or avenues that you can look along or use for photographing wildlife.

Consider building a blind on the edge of a wetland or a woods opening. Hunting-supply companies offer a range of premade, easy-to-set-up blinds, or you can build a permanent one using two-by-fours and plywood. Box blinds can be placed on the ground or elevated on posts to allow better visibility. Plans are widely available on the internet. Use your blind for hunting, watching wildlife, or photographing animals.

A good pair of binoculars is an investment that will pay back its cost many times over in satisfaction. Close-focus models let you observe insects such as butterflies and dragonflies, frogs in vernal ponds, small mammals such as voles and shrews, and birds that are right under your nose in thick cover—or focus on a big buck deer or a fox half a hayfield away. Binoculars with impressively good optics can be had for $200 to $400. Learn about your options at All About Birds, the website of the Cornell Lab of Ornithology. If you have a pond or an open wetland, consider a spotting scope. It's tremendously exciting to be able to see ducks up close in their striking spring breeding plumage or a heron or egret stalking the shallows for prey.

WITNESSING FOR WILDLIFE

Keep a journal of your wildlife sightings, recording the animals you meet and where and when you see them. Not only will this activity prompt you to get out and pay attention to the different kinds of wildlife your habitat attracts, but it will also remind you of what's due to take place in natural settings in the weeks and months to come. The Cornell Lab of Ornithology hosts eBird, which offers anyone a means of recording and storing their sightings of birds, along with pertinent observations, photographs, and sound recordings, and making them available to educators, conservationists, and other birders. This powerful electronic tool is helping scientists gain a better understanding of the movements and habitat needs of birds—plus it will help you find the best birding hotspots in your home area, your state, or anyplace you visit when traveling. iNaturalist is a similar citizen science project and online social network active since 2008. An app for mobile devices, Seek by iNaturalist, helps children and families learn about nature and lets them earn badges for observations.

To keep tabs on wildlife using habitat on your land, put up one or a series of battery-powered digital trail cameras. Different models can take still photos or videos and record sound; they rely on removable memory cards. A biologist studying New England cottontails in New York's Hudson River Valley captured this shot of a coyote in a lush summer habitat.
Amanda Cheeseman

As you come to know your land better—and as you discover the different kinds of wildlife it attracts at different times of the day and during different seasons—you can share with your neighbors the ways in which you have chosen to make a home for wildlife. Think about hosting nature group get-togethers. You could give a short talk and then lead a field trip for a local garden club, a high-school class, a municipal conservation commission, or a forest stewardship organization whose goal is to help landowners make well-informed decisions on how to manage their land to promote healthy ecosystems and help wildlife. Contact your state university cooperative extension service to learn of such organizations in your area.

You can help educate your peers, and no doubt they will teach you important things in return. After I rehabilitated my woods road network and restored the two old-field habitats on our land, I hosted a hike for a local naturalist group. I explained how I had rejuvenated our two two-acre fields while saving the apple trees and thornapple shrubs that had sprung up there. We

Spring Fling

In early spring, female woodcock nest on the ground in young forest. Before nesting, they mate with males who establish "singing grounds" in nearby open areas. At dusk and dawn, the males sound buzzing *peent* calls, then fly up into the air and come circling back down, singing an ethereal bubbling song.
© *iStock.com/MikeLane45*

MOVING AS QUIETLY AS I CAN, I THREAD MY WAY through the brush to the Lower Meadow, an old field about 50 yards downslope from our house.

It's evening, and the light is rapidly draining from the sky. The air is chilly and raw.

I stop and stand on the damp ground at the edge of the meadow. I'm all ears. I can hear spring peepers calling from a vernal pool in the upper hayfield; their high-pitched cries sound like sleigh bells in the distance.

A crow caws from somewhere in the darkening woods.

In front of me, the old field is studded with shrubs and carpeted with grasses that were matted down by snow last winter. The grasses appear drab and gray, a bit darker than the sky.

Peent!

The strange buzzy sound emanates from the ground in front of me, and it makes my heart sing. It's a regular rite of spring for me, to go out and listen for the woodcock telling me that they have arrived from their southern wintering grounds, that winter is truly over and spring has finally begun. Here in Vermont, that's usually the last week of March or the first week of April. When I lived in Pennsylvania, most years I'd hear the first woodcock in late February.

Peent!

It is the male woodcock who calls. He only makes that unmusical sound while he's sitting on the ground.

Peent!

I steal in a bit closer, getting to within 25 yards and stopping behind a small fir. From here I can see a patch of sky above the singing ground.

Peent!

He sounds his rasping call repeatedly for several more minutes. Then I hear his wings twittering as he takes off. He flies up, spiraling into the sky. When he reaches the height of his ascent—maybe 75 yards above the ground—I lose sight of him against the darkening firmament. Soon he comes sweeping downward, flying in broad circles, singing out a bubbling, ethereal *p chuck tee chuck chip chip chip chip*. It reminds me of a robin's calling, but the woodcock's song seems wilder and purer.

The bird sweeps over my head, and I glimpse his rapidly beating wings and his long bill. It's just barely light enough for me to make out the woodcock hovering briefly above the ground before pitching down to land in the site from which he took off. Immediately, he sounds that decidedly unmusical call: *Peent!*

A few times I've been lucky enough to spot a hen winging in to land near the calling male. On the ground (and I've seen this only once, through binoculars), the male fans his white-tipped tail and struts like a miniature turkey gobbler. Following mating, the hen will find a spot in suitable young-forest cover nearby to nest. She doesn't really build a nest—it's just a little dip in the leaf litter. The male doesn't help with mundane chores such as incubating eggs or rearing young. He just keeps on with his spring fling, flying and singing at dusk and dawn for another month as he tries to lure in other hens to mate. A male will guard his singing ground jealously, driving off interloping males by flying straight at them while chattering loudly.

Every few years in late summer I get out my brush hog and chew down the trees that keep invading the grasses in the Lower Meadow. I want it to remain a woodcock singing ground. The work has borne dividends. Not a spring goes by that I am not privileged to hear the woodcock's song.

checked out the two habitats on our hike. Elsewhere along our woods roads, a retired plant pathologist informed the dozen other hikers and me about the complexities of beech bark disease (some of our beeches are afflicted) and what that malady may mean for eastern forests. Another sharp-eyed participant, a natural sciences professor at a local college, spotted and quickly identified an obscure and relatively rare shrub on the forest floor: Leatherwood, which puts forth beautiful twinned yellow flowers in springtime and, when not in flower, looks a lot like an invasive honeysuckle. (I was glad to learn this so as not to make the egregious mistake of pulling leatherwoods out by their roots.) Leatherwood is so named because Native Americans used its pliable, tough stems for making baskets. Turns out it's a great shrub for shady spots and small woodland gardens and for planting along stream banks. Deer avoid eating the plant (its foliage must produce some chemical deterrent), and its leaves turn a beautiful lemon-yellow in fall. The caterpillar of an obscure moth, *Leucanthiza dircella* (it doesn't seem to have a common name), is the only insect known to eat this not-very-palatable native shrub.

While I had a great and a memorable time hosting the naturalists—and showing our two meadows to an ecology class from a local high school, as well—I find I get my deepest enjoyment out of venturing out on the land by myself. Like the time I was sitting in a tree stand at dawn on the edge of the Upper Meadow when a coyote came sneaking out of the woods to feed on apples beneath one of the trees I'd saved; a dark gray shape in the dim light, he ate for about 10 minutes, lifting his head every so often, his ears pricked as he listened for danger.

Or the day when I was strolling along on one of our roads, only to have the peace broken by a pair of raccoons squalling and screeching as they mated most vociferously about 20 feet up in a huge, old sugar maple, a tree with an expansive hollow in its massive trunk. Then there was the tiny brown creeper preening the bark of a maple. The equally small winter wren, its tail bobbing emphatically as it scolded me. And the time I was dubbing along on one of my sneaks on a gray fall day, rain spitting down, when, out of the corner of my eye, I caught a flash of white: an ermine questing between downed tree branches as it hunted for prey.

So go ahead, get out and about. At dawn and dusk and during the middle of the day. On frozen nights when the moon glints on the snow. On rainy November afternoons as the light is shutting down, when you can go along as quietly as a coyote. Enjoy the fruits of your labor and welcome the wildlife you have helped.

Index

Italicized page numbers indicate illustrations.

About the Author

Author photo by Garet Nelson

Charles Fergus has published 16 books on nature and wildlife, including *The Wingless Crow* and *Thornapples: The Naturalist's Year* (collections of nature essays), *Wildlife of Pennsylvania and the Northeast, Wildlife of Virginia and Maryland, Trees of Pennsylvania and the Northeast, Trees of New England, Common Edible and Poisonous Mushrooms of the Northeast*, and two books in Stackpole's Wild Guides series: *Bears* and *Turtles*. A communications consultant for the Wildlife Management Institute, he helps produce publications, displays, and Best Habitat Management Practices guidebooks, and he handles three websites about wildlife that need young forest habitat. Fergus has shared a nature column, "Crossings," in *Pennsylvania Game News* magazine for the past 15 years; before that, he wrote the popular "Thornapples" column for the same magazine. In addition to his books, he has written about nature and wildlife for publications including *Country Journal, Highlights for Children, Audubon*, and the *New York Times*.

Fergus is also a novelist and a writer of historical mysteries, including *A Stranger Here Below*. See www.charlesfergus.com.